W9-ACM-692

SPACE 50

PIERS BIZONY

Smithsonian Books

SPACE 50

HarperCollins books may be purchased for educational, business, or sales promotional use. For information please write: Special Markets Department, HarperCollins Publishers Inc., 10 East 53rd Street, New York, NY 10022.

Published 2006 in the United States
of America by Smithsonian Books
In association with HarperCollins Publishers.

The Library of Congress Cataloging In Publication Data Has Been Applied For.

ISBN-10: 006089010X
ISBN-13: 9780060890100

SPACE 50 produced for Smithsonian Books by Here+There

Art Director: Caz Hildebrand
Designer: Mark Paton
Copy Editor: James Kingsland
Indexer: Vanessa Bird
www.hereandtheregroup.com

Printed in Singapore by Imago

"First, inevitably, comes the idea, the fantasy, the fairy tale. Then, scientific calculation. Ultimately, fulfillment crowns the dream."

Russian pioneer space theorist
Konstantin Tsiolkovsky, 1926.

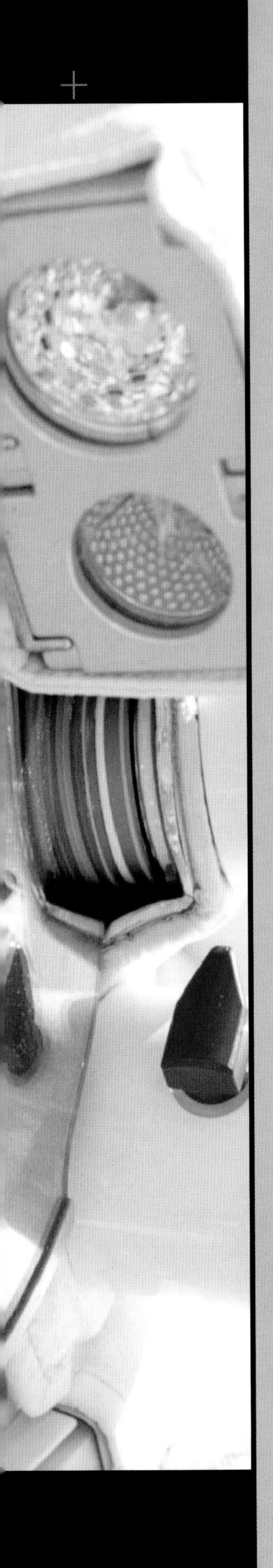

CONTENTS

Foreword

At the beginning of the twentieth century, almost everyone believed that heavier-than-air flight was impossible. The celebrated American astronomer, Simon Newcomb, claimed to have proved this beyond all reasonable doubt. As a result, when the Wright brothers flew in 1903, no one took any notice for over five years—the report must obviously be a hoax! Not until the Wrights went to France, and created a sensation there, did America catch up with the news.

When the Space Age opened on 4 October 1957 with the launch of Sputnik 1, it took about five minutes—not five years—for the world to realize what had happened. Even though I had been writing and speaking about this possibility for years before, I have vivid memories of exactly when I heard about it myself.

I was in Barcelona that week for the 8th International Astronautical Congress. We had retired to our hotel rooms after a busy day of presentations by the time the news broke. In fact, I was awakened by reporters seeking authoritative comment on the Soviet achievement. For the next few days, the Barcelona Congress became the scene of much animated discussion about what the United States could do to regain some of its scientific prestige. While manned spaceflight and moon landings were widely speculated about, many still harboured doubts about an American lead in space. One delegate, noticing that there were 23 American and five Soviet papers at the Congress, remarked that the Americans talked a lot about spaceflight, but the Russians just went and did it.

So much has happened since that momentous week, and many dreams of science fiction writers have come to pass—including a few of my own. We have been to the moon, and created a permanent space station. Nearly five hundred men and women from dozens of nations have been to earth orbit. Unmanned space probes have either landed on, or flown past, all planets excepting Pluto, returning a wealth of information and thousands of stunning images. Orbiting space telescopes are investigating the deepest reaches of our universe, and their results are available for all of us to access on the World Wide Web (which was inspired by one of my short stories).

Closer home, the communications satellite I invented 60 years ago (through my paper in *Wireless World*, October 1945) has changed our world beyond recognition in less than two generations. It has helped create an inter-connected and incessantly chatting global village.

We have certainly come a long way since we "space cadets" of the British Interplanetary Society were desperately trying to promote the idea of space travel back in the 1940s – no, 1930s!

The first half century of space travel has recorded triumphs and setbacks. No one expected that the first chapter of lunar exploration would end after only a dozen men from one nation had walked upon the Moon. Neither did anyone imagine, in those heady days of Apollo, that the solar system would be lost—at least for a while—in the paddy fields of Vietnam. But in the long perspective of history, a few odd decades' delay does not really matter. Sooner or later, humans will return to space.

This is not the first time that our ambition had outrun technology. In the Antarctic summer of 1911-12, ten men reached the South Pole, and five returned. They used only the most primitive of tools and energy sources—snowshoes, dog sleds, their own muscles. Once the pole had been attained, it was abandoned for nearly half a century. And then, in the International Geophysical Year (1957-58), humans came back with all the resources of modern technology—and they stayed. For over 40 years now, summer and winter, men and women have been living and working at the South Pole.

So it will be with the moon. When we return in the coming decades, it will be in vehicles that make the Saturn V rocket look like a clumsy, inefficient dinosaur of the early Space Age (which it was). And this time, we will stay—and slowly extend onward to the planets, beginning with Mars.

Piers Bizony's thoughtfully written and beautifully illustrated book gives us a chance to pause for a moment and reflect on the past 50 years of space exploration, and the next half century that stretches ahead of us, laden with opportunity. Already there have been many epic achievements—and some terrible tragedies. Some of the most wonderful technologies have been realized, and others remain unbuilt, or abandoned too soon. As this book reminds us, space exploration was born in a time of Cold War, and it may yet die in some other, hotter conflict.

I believe that the Golden Age of space travel is still ahead of us. Before this decade is out, fee-paying passengers will be experiencing sub-orbital flights aboard privately funded passenger vehicles, built by a new generation of engineer-entrepreneurs with an unstoppable passion for space. And over the next 50 years, thousands will gain access to the orbital realm —and then, to the Moon and beyond.

There is at least one idea that will ultimately make space transport cheap and affordable to ordinary people: the Space Elevator. It has been around for 40 years, and is now being taken seriously: space agencies are beginning to invest money and effort in developing it.

As its most enthusiastic promoter, I have often been asked when I think the first Space Elevator might be built. My answer has always been: about 50 years after everyone has stopped laughing.

No one is laughing any longer.

Sir Arthur C. Clarke
Colombo, Sri Lanka
June 2006

01 KINGDOMS OF THE ROCKET MAKERS

"I don't want to exaggerate but I'd say we gawked at what he had to show us, as if we were a bunch of sheep seeing a new gate for the first time. Korolev took us on a tour of the launching pad and tried to explain to us how the rocket worked. We didn't believe it could fly. We were like peasants in a marketplace, walking around the rocket, touching it, tapping it to see if it was sturdy enough."

Former Soviet Premier Nikita Khrushchev, 1970

KINGDOMS OF THE ROCKET MAKERS

> Half a century ago, two brilliant men led the epic space race between the Soviet Union and the United States. Both men had known the worst cruelties of the twentieth century. One narrowly avoided death in a prison camp, while the other may have been complicit in running one. Both men built rocket weapons of appalling destructive power, even as they propelled us into the cosmos. Both believed in a shining vision of human progress and ascent to the stars, yet they created a technology that could just as easily have brought an end to all such hopes. The contradictions personified by these two individuals represent perfectly the two faces of this new adventure. The development of space rockets highlighted the best and the worst of our desires. They satisfied our yearning to explore a heavenly realm in peace, but their design was inspired by earthly wars.

The man responsible for the early triumphs of the Soviet space effort was a shadowy figure. Most of his colleagues referred to him as "The King" or "The Big Boss", or else they affectionately used his first two initials, "S.P." His full name was never spoken in public because the Kremlin had declared his identity a state secret. In the many press and radio reports of Soviet rocket achievements published and broadcast over the years, he was referred to as "The Chief Designer". This deliberate aura of mystery enhanced the nervousness that he provoked among the U.S. space community. Born in 1907 in the Ukraine and educated in Moscow, Sergei Pavlovich Korolev began his career as an aircraft designer in 1930 before developing a fascination for rockets. At first he saw them as a useful power source for his aircraft, but by the late 1930s he recognized that they had potential as vehicles in their own right. He began to think about space.

A 1963 wind tunnel model of a Centaur rocket is tested to check how well it can ease excess hydrogen fuel pressure during flight.

Before World War II Russian military strategists saw some potential in rockets, at least on a small scale as battlefield weapons. Marshal Mikhail Tukhachevsky sponsored a new research center, the Gas Dynamics Laboratory, which was hidden away behind the heavy ramparts of the Petropavlovskaya Fortress in St. Petersberg (scorchmarks in the masonry can still be seen), while another lab in central Moscow, the Reaction Propulsion Laboratory, worked on similar problems. From these parallel efforts Korolev's main rival, Valentin Glushko, emerged as the most promising designer of rocket thrust chambers and fuel pumps, while Korolev thought in broader terms about how to combine the engines with fuel tanks, guidance equipment and a payload. He had in mind large rocket vehicles that could perform useful work: delivering bombs, taking weather measurements in the high atmosphere and, one day, exploring space.

Tukhachevsky was mainly interested in winged rocket bombs and other armaments for the Red Army. In 1933 he tried to consolidate the various rocket programs into a unified military program. Unfortunately Joseph Stalin was terrified of such independent-minded soldiers and by 1937 had initiated a wide-ranging purge of the officer class as part of his regime of terror throughout all levels of Soviet society. Tukhachevsky was arrested on June 11 and shot dead that same night. All the rocket engineers he had assembled and sponsored were arrested and charged with treason. Korolev was dragged away on June 27, 1938, and condemned to ten years' hard labor in Siberia: essentially a death sentence.

In June 1941 the invading Nazis scored devastating victories against a thoroughly unprepared Red Army. Stalin soon had cause to regret his earlier purge of the officer corps. His remaining commanders, perhaps understandably, were more adept at personal survival than attending to the broader needs of military strategy. Korolev and many other vital engineering personnel were released from their prison camps and taken back to the major industrial centers to work in aircraft and weapons factories, although still under guard. Towards the end of the war Korolev was freed from captivity with his reputation partially restored, and in September 1945 he was allowed to venture into the crumbling German heartland in search of any remnants of the Nazis' brilliant but controversial V-2 rocket bomb program that the British and American forces hadn't already taken away. Throughout the 1950s, with incredible energy and determination, he developed an increasingly sophisticated range of rockets and missiles, while Glushko designed some of the most effective propulsion engines the world had ever seen.

For a man who had survived a Siberian labor camp, mere administrative battles with competitors or unhelpful officials at the Kremlin must have seemed relatively easy, particularly under the far less oppressive post-Stalinist regime of Nikita Khrushchev throughout the late 1950s and early 1960s. Korolev was a driven man, and he established a network of influence far more complex and subtle than anything his rivals in the aerospace sector could muster. By 1956 he controlled his own industrial empire, the heart of which was a secret factory in Kaliningrad, northeast of Moscow, known only as the Special Design Bureau-1 (OKB-1). Korolev was the absolute ruler here, although he was answerable at the Kremlin level to the Defense Ministry under Marshal Ustinov and also to the vaguely named Ministry of General Machine Building. In this context "General" was a codeword for rockets.

Yuri Mazzhorin, Korolev's designer of guidance systems, remembers his boss with fondness. "He was a great man. You'd think his time in prison would have broken his spirit, but when I first met him in Germany when we were investigating the V-2 weapons, he was a king, a strong-willed purposeful person who knew exactly what he wanted. He shouted and swore at you, but never insulted you. He would listen to what you had to say. The truth is, everybody loved him."

Almost everybody. Valentin Glushko, an equally driven personality, operated out of his own specialist design bureau, OKB-456 in Khimki, northwest of Moscow, and dangerously near Korolev's OKB-1. As long as his engines were fitted into Korolev's rockets, the two men avoided outright confrontation, but these two giants of Soviet rocketry did not get along. The tension between them undoubtedly dated back to the summer of 1938 when, for some reason, Glushko was punished with eight months of relatively mild house arrest while Korolev was sent to a prison camp. Glushko seems to have betrayed some of his colleagues while Korolev kept a costly silence. The Chief Designer's illicit affair with Glushko's sister-in-law Ketovania Sarkisova seems only to have heightened the tension between the two men.

Mikhail Yangel was another rival, developing missiles strictly for military use from his bureau in Dniepropetrovsk in the Ukraine. The fourth major figure in Soviet rocket development, Vladimir Chelomei, had the presence of mind to hire Nikita Khrushchev's son Sergei as an engineer. (Chelomei's OKB-52 design bureau eventually created a rocket called Proton, an unmanned booster capable of lifting large payloads such as space station modules. Just as Korolev's family of rockets are still in use, so are modernized versions of Chelomei's.)

But by far the most serious challenge to Korolev's autonomy came from high-ranking military officers based in and around the Kremlin and the Ministry of Defense who were concerned that his ambitions for space would interfere with the development of weapons systems. He appeased them by designing the R-7 missile, a machine capable of dropping a nuclear bomb into the American heartland. Russia's early warheads were large and heavy, and Korolev knew that if his R-7 could lift these, it could lift anything, including spaceships.

Mstislav Keldysh, a senior member of Russia's prestigious Academy of Sciences and one of the Chief Designer's most powerful allies, supported the idea of space missions and scientific experiments in orbit. He was an expert on the mathematics of missile and rocket trajectories, and had established a power base in Moscow centered on his custom-built computing facility, sponsored by the Academy. While Korolev built the rockets, Keldysh plotted the routes they would fly.

EDA
EDA

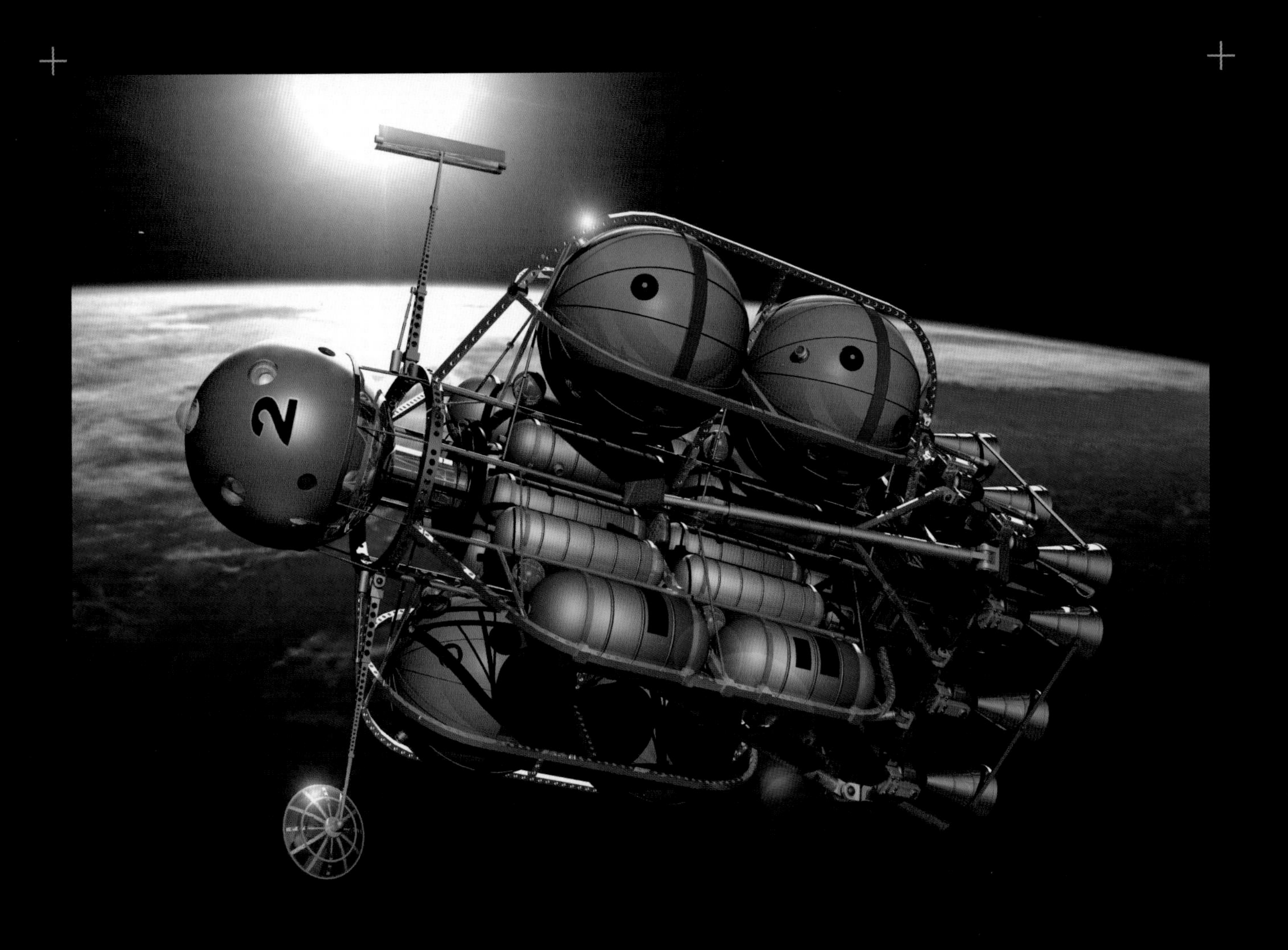

Opposite, King Peter of Yugoslavia and his son Prince Alexander visit a space-related exhbibition in London during 1953. Children at that time were fascinated by the promise of space. What new dreams might inspire today's young people?

A modern computer graphics rendition of von Braun's original moonship proposal by digital artist Terry Sunday brings a half-century old concept vividly back to life. NASA now plans to revisit the moon in the coming decade, using lunar landers that have a startlingly similar appearance to the spacecraft depicted above.

"I'm sure we would not have had men on the moon if it had not been for Wells and Verne and the people who write about this and made people think about it. I'm rather proud of the fact that I know several astronauts who became astronauts through reading my books."

Arthur C. Clarke, Address to US Congress, 1975

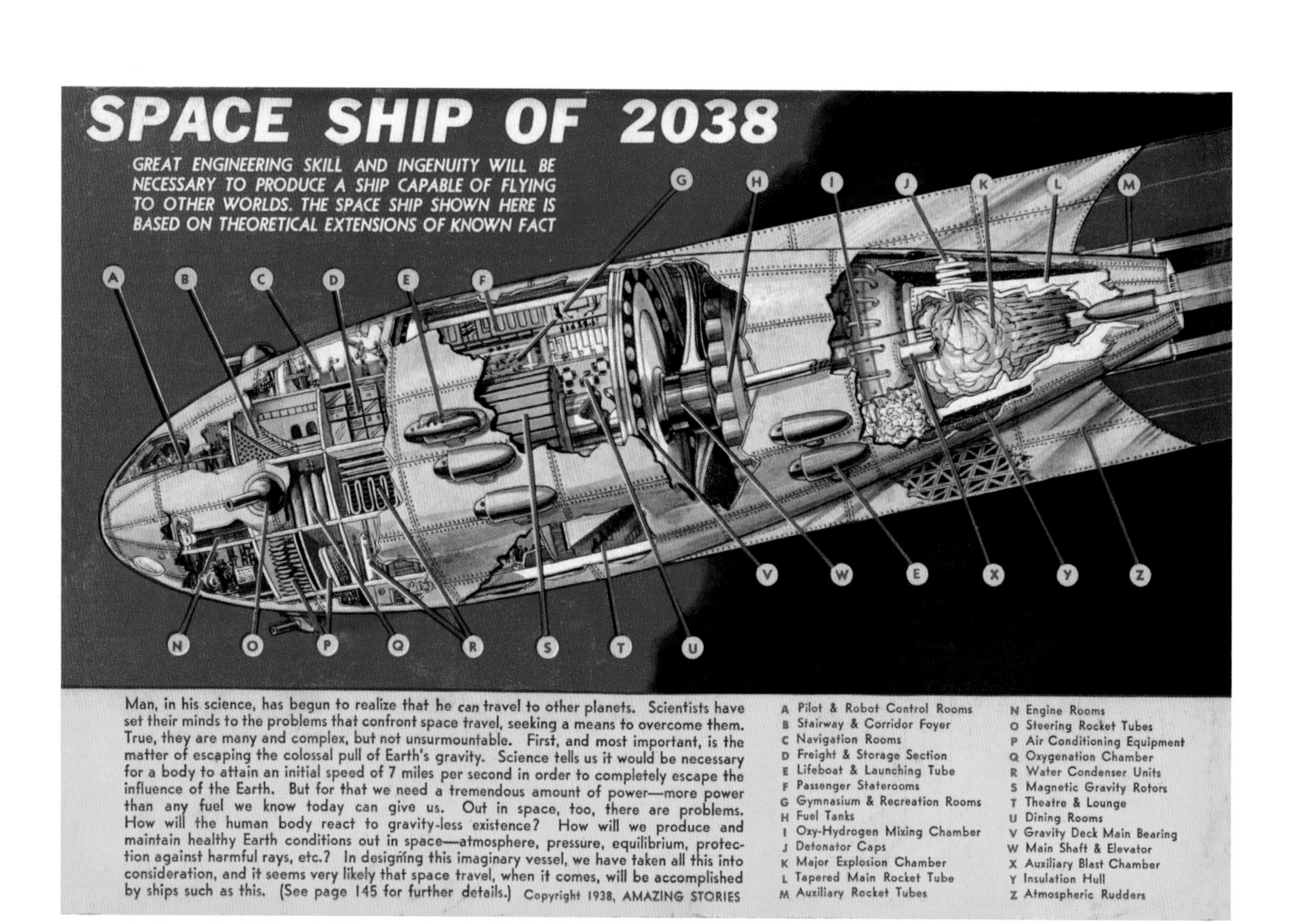

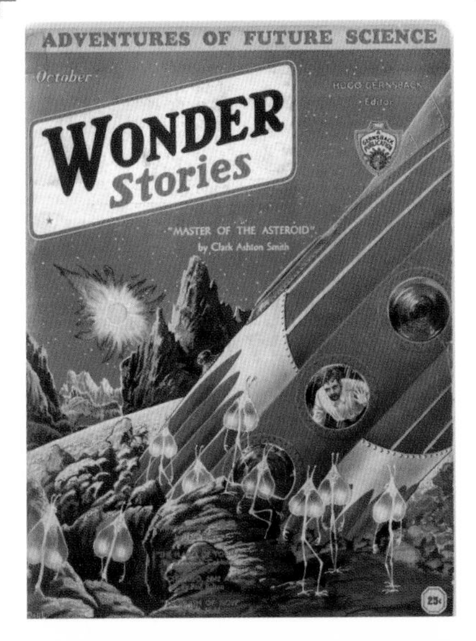

Science fiction novels and comic strips were essential spurs to the imagination prior to the arrival of real space hardware. The engineers and politicians in the Cold War era did not sudddenly invent the glamor of space travel out of nowhere. It had already been planted in the popular imagination by writers and artists.

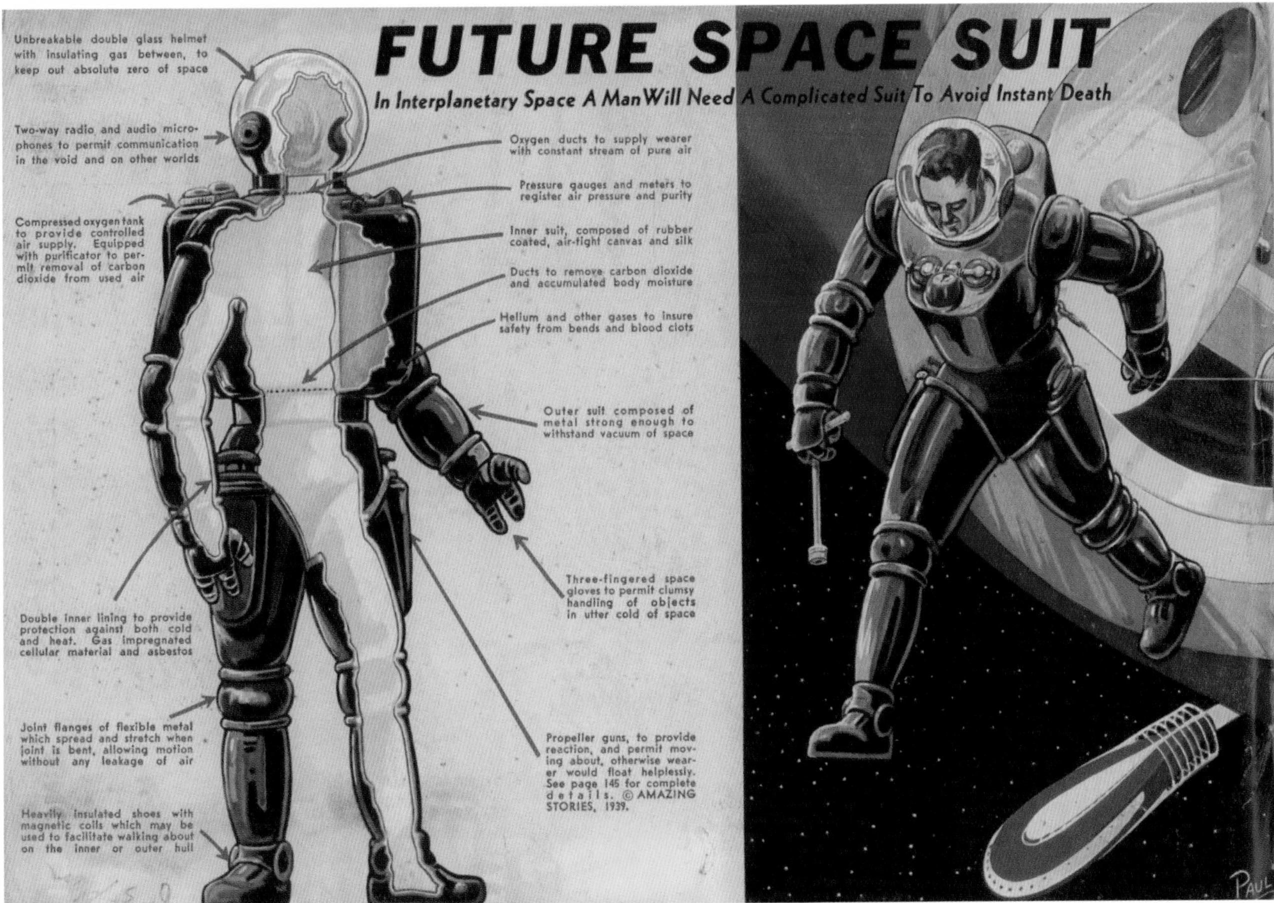

"Korolev was a great man. You'd think his time in prison would have broken his spirit, but when I first met him in Germany when we were investigating the V-2 weapons, he was a king, a strong-willed purposeful person who knew exactly what he wanted. He shouted and swore at you, but he never insulted you. And he would always listen to what you had to say. The truth is, everybody loved him."

Yuri Mazzhorin,
former colleague, 1997

Korolev would have recognized this view of a modern Soyuz rocket heading towards its launchpad by rail. Much of the hardware is similar to the original designs from half a century ago. Russian rockets may be of an old design, but they are rugged and reliable.

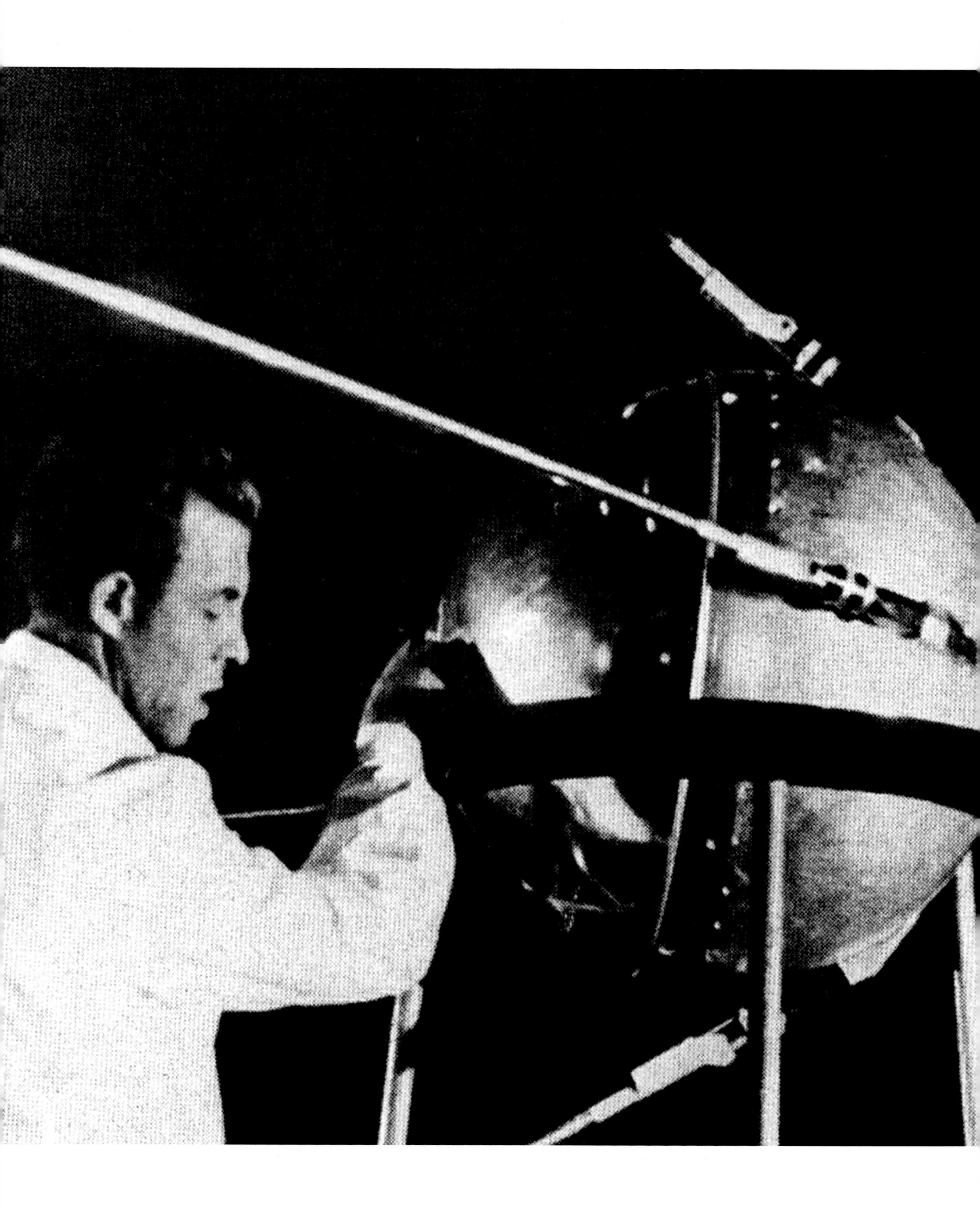

A promotional film frame shows a technician checking Sputnik prior to launch. The scene was staged after the mission.

The most irritating problem for Korolev was that he still had to rely on the Soviet Army's unintelligent cooperation, because his work with rockets and missiles was so intimately linked with the military. He could subvert military equipment to his own ends, but he couldn't make the jealous generals vanish. However, if one of them blocked him he was quite unafraid to treat the man as an inferior.

A senior engineer in Korolev's bureau, Oleg Ivanovsky, was amazed by his boss's nerve. "On one occasion a very high-ranking commander refused access to an important radio link during a flight. Korolev spoke to him on an open phone line and shouted, 'You don't know how to do your job! Give me the link, or I'll have you demoted to sergeant!' We were amazed that he could be so insolent to a superior."

First Secretary Nikita Khrushchev and his colleagues in the Politburo could be very supportive towards Korolev when it suited them, although they weren't particularly concerned with the subtleties of rocket hardware. The technology fascinated them more in its glamor and potential political impact than in its hard engineering. In the summer of 1955, Korolev asked senior members of the Politburo to inspect his work, as Khrushchev recalled in his memoirs: "I don't want to exaggerate but I'd say we gawked at what he had to show us, as if we were a bunch of sheep seeing a new gate for the first time. Korolev took us on a tour of the launching pad and tried to explain to us how the rocket worked. We didn't believe it could fly. We were like peasants in a marketplace, walking around the rocket, touching it, tapping it to see if it was sturdy enough."

Korolev's colleague Sergei Belotserkovsky (responsible for cosmonauts' academic training) sums up the Politburo's attitude to space projects. "The top people's opinion about Korolev was purely that of consumers. For as long as he was indispensable, for as long as they needed him to develop missiles as a shield for the motherland, he was allowed to do whatever was necessary, but all his space research had to follow on the back of the military work. The point is that he eventually launched his spaceships and cosmonauts on the very same missiles."

His principal workhorse was the dual-purpose R-7, or Semyorka (Little Seven) as it was affectionately known by the men who built it or flew on it. Fuelled with kerosene and liquid oxygen, and incorporating four drop-away side boosters, this was the world's first operational intercontinental ballistic missile (ICBM). It was powered by Valentin Glushko's undeniably well-designed engines. Setting aside their rivalries for the moment, Glushko and Korolev between them created a machine that is still in use today, modernized over the years, but basically the same as it was more than half a century ago when it was first designed.

The first two launches of the R-7 failed, but on August 3, 1957, it flew successfully in a simulated ICBM trajectory. It began its career as a space launcher just two months later on October 4 when it launched Sputnik, the world's first artificial satellite.

"Repeatedly, I saw President Eisenhower angered by the excesses, both in text and in advertising, of the aerospace-electronics press, which advocated bigger and better weapons to meet an ever bigger and better Soviet threat that they conjured up. I remember the shrill, hard campaigns by a few corporate lobbyists in support of their companies. The Sputnik panic was being used to support an orgy of technological fantasies."

Former White House Science Advisor James R. Killian, 1977

A Soyuz rocket is rolled towards the launchpad in 2005, preparatory to a space station mission. This rear view shows almost exactly the same kind of booster hardware that launched Sputnik into orbit, heralding the space age.

Were the epic events that lay ahead somehow inevitable? Was it only a matter of time before we headed for the moon? Had it really taken just this one particularly driven individual, Sergei Korolev, to bring in the entire Space Age? Andy Aldrin, son of Apollo 11's Buzz Aldrin and Vice President of Boeing Launch Services, has studied the history of the Soviet rocket effort. "I do not think Russia would have beaten the United States into space without the strategies that Korolev and his team developed. He was the primary motivator. Beyond a small group working with him, there was not any broad support for space exploration until after Sputnik. Neither the political, military, nor industrial leadership had any desire to go into space. In many ways Korolev tricked them into doing it."

When we tell stories about the past, we have a tendency to focus on a few key individuals. History books are replete with kings whose biographies are an easily digestible shorthand for how entire kingdoms behaved. We read about the tactics of generals who led armies because we cannot sensibly absorb the stories of every soldier and every last skirmish on the battlefield. We admire engineers who designed great bridges or mighty rockets, yet who did not personally rivet them together or put fuel in their tanks. Stories of the Space Age cannot help but fall into this trap, for by the mid-1960s, for a brief few years at least, more than a million people on both sides of the space race were working in or around the field of rocketry. And so we come to Korolev's great rival, another Space Age "king" who stands for a much greater human story than just his own.

Von Braun

> Wernher von Braun is probably the only rocket engineer out of countless thousands whom the general public has actually heard of. They know he worked for NASA and designed the giant Saturn booster that took Neil Armstrong and Buzz Aldrin to the moon in July 1969. They know he built V-2 rocket bombs in World War II. Von Braun was the archetype for our image of a "rocket scientist", complete with German accent. He was born in 1912 into an aristocratic German family in Wirsitz, Prussia. The von Brauns relocated to Berlin in 1920 after their original homeland had been ceded to Poland in the wake of the troubled peace treaties of World War I.

The young von Braun became fascinated by science fiction—that literary underdog of pulp magazines and lurid fantasies which has, in fact, transformed the world just as much as more "sophisticated" literature ever has. Jules Verne's 1865 novel *From the Earth to the moon* was a particular inspiration. Like so many space visionaries, von Braun fell in love with rockets as a young boy and chased that one passion relentlessly for the rest of his life. Rockets today are "just" machines for sending satellites and modules into space, but half a century ago they seemed to promise something more to the people who dreamed of building them. Humanity would fly into the cosmos and settle on other planets, where new and perhaps better societies could be created. The rocket mirage was vague, but it was in keeping with the "romance of flight" or the "romance of the open road" that made the airplane and the automobile such objects of desire in the 1920s and 1930s, when the darker potentials of such machines had not yet been revealed to their fullest and most bitter extent.

Von Braun allied himself with a budding organization of rocket enthusiasts known as The Society for Space Travel (Verein für Raumschiffahrt or VfR). The society was formed in 1927 by like-minded rocket visionaries, including Willy Ley and Hermann Oberth, scientifically experienced men with certain valuable skills beyond math, physics and engineering: the ability to write clear and accessible books promoting rocket technology. The VfR built small-scale amateur rockets and quickly developed ideas that attracted interest from the German military. Oberth and von Braun were delighted by the prospect of serious funding, but by 1935 Ley had emigrated to America, sickened by the rise of Nazism in his home country.

In the late 1930s von Braun and his remaining close-knit team began working on the world's first long-range military missiles at Peenemünde on the Baltic coast, under the supervision of General Walter Dornburger (a man also captivated by the romance of space rockets, as well as their military potential). The dream soured quickly. It was not von Braun but the Nazi leadership who eventually came up with the name "Vergeltungswaffe 2", Vengeance Weapon 2, now universally known as the V-2. (An earlier and much smaller flying bomb, unrelated to von Braun's work, had already been designated the V-1.) The majority of V-2 rockets that were launched as weapons (primarily against Britain) were assembled at a huge under-

Wernher von Braun at his office at NASA's Marshall Space Flight Center in Hunstville, Alabama, surrounded by models of his rocket creations.

ground facility, Mittelwerke, at the southern end of the Harz mountain range, a few miles outside the city of Nordhausen. A vast warren of tunnels was built by thousands of concentration camp prisoners working in the most appalling conditions. How much did von Braun know about these abuses? This is an important and as-yet unresolved question about a man whose life, in so many other respects, was inspirational.

Dr. Michael Neufeld of the Smithsonian Institution has identified documents suggesting that von Braun was perfectly aware of the slave labor used to construct his rockets. "The more time goes by, the harder it is to take the accepted view that he was just a disinterested bystander in all this. He knew all the worst that was going on." The first clue lies within the most innocent of documents. Von Braun's personal pilot's log shows the numerous dates when he flew from Peenemünde to Nordhausen. Since most of those dates lie within the period during which the atrocities were at their most obvious, the inference is clear. More specifically, an internal memo from von Braun's deputy Arthur Rudolph, dated April 12, 1943, says:

"The use of detainee labor has worked well [at the Heinkel aircraft factory]. The mixture of multiple nationalities helps deter sabotage, and there are many advantages over the earlier use of foreigners, because the non-work-related tasks are taken over by the SS."

That casual phrase "non-work-related tasks" was a euphemism for the SS operation of concentration camp facilities. The Buchenwald camp was exploited as a source of labor for Nordhausen. On August 15, 1944, von Braun wrote to one of his V-2 production managers, Albin Sawatzki:

"During my last visit to Nordhausen, you proposed to me that we use the technical education of detainees available to you and Buchenwald to tackle additional jobs. I immediately looked into your proposal by going to Buchenwald to seek qualified detainees. I have arranged their transfer to Nordhausen with the Buchenwald camp commandant."

But there may be more to von Braun's story than meets the eye. Do we not have to imagine ourselves in his shoes before reaching a moral judgment? Was it wrong to fight for his homeland, or to develop weapons for the ruling elite of his own country? Was he young and naïve, and swept away by terrible events over which he had little control? Was there any option to turn away from his work once the Nazi authorities had identified its value? It has even been suggested that he drew more technicians out of Buchenwald than he strictly needed, perhaps in an effort to save them. There is no firm evidence either way. But very many people who knew von Braun considered him a decent and honorable man.

Historians still debate the significance of von Braun's acceptance of an officer's rank in the Waffen-SS. By the end of the war he had been promoted to Major, although there is no suggestion that he welcomed or encouraged this. Certainly he was advised by colleagues that he had better accept the uniform, for surely he and his work would be endangered if he did not. His arrest in February 1944 also poses a challenge for historians. SS chief Heinrich Himmler had told von Braun that he should cooperate in transferring the rocket program from the army to the SS. "I trust you realize that your rocket has ceased to be an engineer's toy? The German people are eagerly waiting for it."

Von Braun politely suggested that his arrangements with the army were quite adequate. It is probable that he did not welcome being hijacked by the SS. But this is essentially what happened. Shortly after that chilly meeting with Himmler, he was arrested and charged with sabotage. His crime? Daring to talk about the possibilities of spaceflight instead of focusing on his

Right, a von Braun sketch for a winged space plane on an even larger winged booster. This concept inspired NASA's shuttle craft, although a flying booster stage proved too expensive to build. Opposite, in an extravagant artwork for *Collier's* magazine by Fred Freeman, a fleet of landing ships and giant winged gliders heads for Mars.

military work. More specifically, he was caught up in the struggle between the army and the SS for control over the rockets. Inevitably the SS got its way, and its hellish influence eventually dominated every aspect of the rocket program. Von Braun was released after a fortnight and continued with his work. But he knew that it would only be a matter of time before the defeat of the Nazi regime. What, then, would happen to him and his rockets?

In the last months of the war, British, American and Russian intelligence teams scoured the devastated German heartland for any remains of the V-2. It was, after all, the world's first guided missile and it would define the future balance of strategic power. Competition among the teams was ruthless, even though they were supposedly allies. Von Braun decided that the American rocket hunters were his best hope, because they would probably employ him. The British couldn't afford him and the Russians might shoot him. He staged a brilliant escape for himself and his closest colleagues under the noses of SS squads, who were by now indiscriminately killing "disloyal" Germans.

In fact the Russians might not have shot him after all. A senior colleague, Helmut Grottrup, and dozens of other V-2 staff were brought to the Soviet Union under quite respectful conditions and debriefed for every last scrap of information. They believed that it was only a matter of time before they would be put to work building rockets, but in the event, after Korolev and his Russian experts had learned everything they could, Grottrup and his people were repatriated to Germany.

For a while it seemed as if von Braun had made the wrong choice of captor. There followed a five-year sojourn at Fort Bliss in Texas, where he and his team briefed skeptical army personnel about rocketry. They were able to conduct V-2 firing tests in the White Sands desert of New Mexico, but essentially von Braun was frustrated by the U.S. government's lack of interest in him, or even in rockets. Clearly as a former Nazi collaborator he was an embarrassment as well as a potential asset. When he and his team were transferred to their next home at the U.S. Army's Redstone Arsenal in Huntsville, Alabama, the workload became more meaningful. As part of the Army Ballistic Missile Agency (ABMA), one of their first tasks was to create the Redstone medium-range missile, named in honor of the arsenal that hosted most of its development. Von Braun was also given the chance to display his undeniable charm on the public stage. He became a campaigner for space, tirelessly working the local Chambers of Commerce, the Rotary clubs and any other venues that would host him. By sheer force of personality, he shed his past and took up the role he had dreamed of ever since he was a boy: as the champion of our ascent into space. Perhaps he reinvented himself morally too, for his speeches and presentations throughout the rest of his life were informed not just by his abiding love of rockets, but also by a passionate belief in human values.

Fred Freeman

Selling the Space Age

> One of the most important presentations von Braun ever made was in a series of articles for *Collier's*, a popular illustrated magazine. Between 1952 and 1954 seven major features on space were published, including descriptions of a lunar colony and a mission to Mars, based intimately on von Braun's technical engineering descriptions, yet phrased in a language anyone could understand. The hardware was brought vividly to life by illustrators Chesley Bonestell, Rolf Klep and Fred Freeman. There was a giant wheel-shaped space station gently turning on its axis, its crew enjoying the artificial gravity generated by the rotation. A painting by Bonestell showed the station attended by winged rocket planes, while in the foreground three huge landing ships are prepared for their trip to the moon "within the next 25 years". Willy Ley, now a successful space popularizer in his own right, was von Braun's major partner in creating the articles, along with many follow-up books. Whatever the divisions of the past, rocketry continued to bind these men as friends and colleagues.

Most of these dreams had been familiar to rocket visionaries and science fiction enthusiasts since the 1930s, but the *Collier's* articles represented perhaps the first time the general public had been asked seriously to consider rocket ships, space stations and trips to the moon as elements of national policy. The magazine sold three million copies a month. As a family title, it would have been read by perhaps fifteen million people. Space was no longer a dream: it was something for taxpayers to think about in earnest.

There followed a succession of popular books and television shows. There were three TV films from Walt Disney, in which von Braun himself featured. The first, *Man in Space*, aired on ABC on March 9, 1955. The second, *Man and the Moon*, aired the same year. The final film, *Mars and Beyond*, was televized in late 1957 just a few weeks after Sputnik had made such dreams seem that much closer. In those simpler days before myriad cable and satellite channels split audiences into hundreds of niche markets, many millions of people would have watched the shows. Von Braun became a trusted household name, and his idealistic, visionary ideas beguiled a generation. Undoubtedly he helped to prepare Americans to accept the forthcoming (and very great) expenses of the Space Age.

America's initial response to Sputnik was an ill-fated Naval Research Laboratory project called Vanguard. On December 6, 1957, it blew apart on the launchpad. "KAPUTNIK!" "STAYPUTNIK!" and "FLOPNIK!" the newspapers jeered. Sensitive to recent World War II history, the Eisenhower administration had been unwilling to allow von Braun to launch a satellite and had chosen Vanguard as America's first space project, supposedly for its technical elegance. In the wake of Vanguard's failure, Eisenhower turned unapologetically now to the U.S. Army missile team at the Redstone Arsenal and their brilliant but somewhat embarrassing principal asset, von Braun. He was ready in just eight weeks. His Jupiter-C rocket (essentially a Redstone missile with a few modifications) carried the tiny satellite Explorer I into orbit on January 31, 1958. Soviet leader Khrushchev disparaged it as

"a grapefruit" because it weighed only 14 kg against Sputnik's 80 kg, but Explorer I immediately made one of the most important scientific discoveries of the twentieth century when Dr. James Van Allen's instruments detected radiation belts around the earth.

Even though they had dreamed of this moment for so long, von Braun and his colleagues were taken aback by the reaction to this success. A press conference at the National Academy of Sciences in Washington D.C. was jam-packed with eager journalists. In fact the Soviet authorities had also been surprised by the strength of feeling generated by Sputnik. It wasn't just politicians but people who spurred the Space Age—for if civilian populations had not been moved by this adventure, then surely there would have been no point in any government promoting it.

Korolev had not, of course, been allowed to "sell" his message directly to the Russian people. Their heartfelt joy at the success of Sputnik was tied to the sense of promise it delivered—a promise that the Communist ideal might actually be possible, despite all the hardships that this colossal social experiment had yielded so far. Soviet planners dreamed of taming Nature in the service of Man. What was Sputnik but a promise that the stars themselves might similarly be tamed? And if the Russian people's long-held religious faith had been denied by the earthly politics of Marx, Lenin and Stalin, then it was free, now, to express itself in a different and more materialist kind of language. What were rockets but new cathedrals? What was Sputnik if not a polished metal prayer? What was the cosmos if not a kind of heaven that Russians could now dream of reaching? The tensions that had existed between the Soviet State and the Russian Orthodox Church briefly found respite among the rockets.

For his part, von Braun hitched his wagon to an equally compelling American dream: the frontier spirit of the Old West. He and many others adapted the country's romantic self-image of its origins—Columbus, the Mayflower, the pioneer trail. A small fleet of ships sets sail across an uncharted ocean to discover new territory; a settlement is established at the point of landing; a settlement becomes a town. In time, other ships arrive carrying families of settlers to populate the new territories. Inspired by von Braun, space visionaries talked in terms of "conquering the Sun's empire" and "colonizing the space frontier". Always, their plans were worked out with exact attention to fuel weights, rocket thrusts, orbital heights and speeds. As the renowned futurologist and space writer Arthur C. Clarke has pointed out, "No achievement in human affairs was ever so well documented before the fact as space travel."

The flaw in the plan was simple: where was the trade? Ships could be sent into space, but what could they be expected to bring back as a return on the huge investment of launching them? This fundamental

Above, Walt Disney, left and Wernher von Braun, right.

A delta-winged rocket model is readied for an aerodynamic test in 1950 opposite.

America's first satellite, Explorer 1, undergoing final checks at the top of the launch gallery, and opposite, Jet Propulsion Laboratory (JPL) director William Pickering, radiation scientist James Van Allen and Wernher von Braun celebrate the successful mission by holding aloft a full-scale replica of the satellite.

economic problem could not be wished away by any amount of romantic language. *Collier's* demonstrated even to the most skeptical audience that spaceflight was possible, and that there were many good scientific and spiritual reasons for initiating some kind of program; but there seemed no obvious way to justify spending the vast sums involved in building a huge space station or a fleet of moon ferries, let alone ships bound for Mars.

In the event, politics rather than commerce sent humanity into space. Von Braun might have been surprised to discover that giant rockets would head for the moon eight years sooner than his *Collier's* deadline. In the wake of Sputnik, the economic arguments against building his dream became strangely inverted, so that the drawbacks became the justifications. By accepting the huge expense of these abstract adventures without prospect of a sensible financial return, Russia and America set out to show what their rival economies and societies might be capable of down on earth—how much *more* they could do than merely running their domestic economies or equipping their armies. It was a battle of economics as much as technology; and only one side could truly expect to win.

The Birth of NASA

Popular and press reaction in America to the launch of Sputnik had been intense and largely fearful. President Eisenhower could not understand why such a tiny and harmless satellite was causing so much fuss. He had little faith in space, and was convinced that this "foolish" fascination would fade away. But he was wrong. His most ardent critic was the Democrat senator Lyndon Johnson, who made a series of influential speeches in support of an expanded rocket program. "First in space means first in everything," he urged.

His audience needed little encouragement.

In the summer of 1958, America's disorganized space projects were pulled together under a new federal authority, the National Aeronautics and Space Administration (NASA), derived from a relatively modest rocket program at The National Advisory Council on Aeronautics (NACA), based in Langley, Virginia. Much of the new agency's strength stemmed from Johnson's legislative skills. Jealous officials at the Pentagon were taken by surprise. NACA had apparently done no more than change one letter in its acronym, yet suddenly it expanded into a powerful (and, at Eisenhower's insistence, nominally civilian) new empire, under the leadership of Keith Glennan from the Case Institute of Technology and his deputy Hugh L. Dryden, NACA's old chief. Glennan readied NASA for the challenges ahead—yet, like Eisenhower, he never expected space flight to grow into the gigantic endeavor that it eventually became. And if the Pentagon and the rest of the military establishment had been denied the glamor of publicly acknowledged space activities, they were certainly not hindered in the development of secret ones. On a practical level, unmanned surveillance satellites were more strategically important than NASA's scientific probes. Behind the scenes, as much money—and in some years, considerably more—was spent on the development and launch of top-secret spy satellites, military communications relays and even prototype space weapons than was ever spent by NASA. The Soviets, already secretive enough about their space activities, were doubly taciturn about the scale and nature of their military experiments in orbit.

In truth, most of the space race was to be a matter of covert war, hidden budgets and total security from prying eyes. The pages of the world's newspapers were persuaded to focus instead on the comparatively innocent promise of "space exploration". Over the coming years, NASA would enjoy the headlines while the Pentagon, the Central Intelligence Agency, the National Reconnaissance Office and other, darker agencies kept their vast budgets concealed from public scrutiny and gave out only partially accurate descriptions of their many satellite prototypes. (The International Outer Space Treaty of 1967 eventually banned the placement of nuclear weapons in orbit, but did little to stem the development of satellites capable of defending themselves against attack, or hindering the operations of "enemy" satellites.)

Eisenhower wished he could limit the scale of space activity, fearful that it could only lead to a new arms race. In the last months of his administration, he reluctantly approved a new rocket called 'Saturn', to be designed by Wernher von Braun's team, now absorbed into NASA. Yet there was no payload for this huge booster. The early Saturn program was not specifically targeted at the moon. Within NASA, the belief was growing that a manned lunar voyage might indeed be possible, but Glennan was cautious. Eisenhower made it clear he would never sanction such a project. He was already disheartened at the press and public attention being given to NASA's first, tiny manned capsule, the Mercury. Yet the trend towards human space flight was unstoppable. Korolev had already laid down that challenge, within just a few weeks of Sputnik's launch.

A V-2 rocket, with a smaller needle-shaped second stage, is test-fired at the White Sands missile range in 1949.

Apes and Dogs

Space historian Andy Aldrin is full of admiration for the speed with which Korolev delivered his second rocket triumph. "He and his merry band of rocket engineers tried to go off on a vacation after Sputnik, and they'd rested for about two days when Korolev got a call from Khrushchev. 'Comrade, we need you to come to the Kremlin.' Of course he went, and he sat down with the Soviet leadership, and they said, 'In a month we have the 40th anniversary of the glorious October Socialist Revolution. We want you to put up another satellite that will do something important.' They proposed a satellite that could broadcast the 'Communist Internationale' anthem from space, but Korolev had another idea. He wanted to put a live animal in the satellite, so that he could lay the groundwork for an eventual manned mission. And within a month, from scratch, he and his people completed the spacecraft and launched it."

Sputnik II went up on November 3, carrying the dog Laika. This was a clear indication where the Soviet space effort was heading. Soviet officials played down the fact that Laika's cabin was cramped and overheated. There was no way to bring her down to earth, and no humane system had been installed for putting her out of her misery. The world came to accept that her oxygen had run out after a few days. In fact dehydration and heat stroke killed her. A truly epic achievement—the sending of a living creature into orbit as a herald for future human voyagers—was tarnished by thoughtless cruelty. But the fact remains that Laika set a precedent from which there could be no retreat.

An unoccupied prototype for a man-carrying capsule, the Vostok, launched on May 15, 1960, spiralled out of control in space and was lost. Two dogs, Chaika and Lisichka, were put aboard another Vostok on July 28 after the capsule had been modified accordingly. Now it was the R-7 booster's turn to disappoint its makers. Shortly after launch the rocket blew itself to pieces, dogs and all. Yuri Gagarin, Gherman Titov, Alexei Leonov and other prospective Vostok "cosmonauts" were at the remote Baikonur launch center that day on their first familiarization visit, and they witnessed the launch of the vehicle supposedly designed to carry them safely into space. Titov wryly recalled: "We saw how the rocket could fly. More important, we saw how it could blow up."

It was not yet safe to launch humans. Quite apart from the temperamental rockets there was a profound unease, on both sides of the superpower divide, about how the human body—and indeed, the mind—would respond to the weightlessness, disorientation and isolation of spaceflight. On August 19 two more Russian dogs, Strelka and Belka, were sent into space, their little bodies riddled with heart sensors and other measuring devices.

Laika, above, in her space harness, and Soviet animal scientist Oleg Gazenko holding aloft Strelka and Belka after their safe return from orbit opposite.

"Work with animals is a source of suffering to all of us. We treat them like babies who cannot speak. The more time passes, the more I'm sorry about it. We shouldn't have done it. We did not learn enough from the mission to justify the death of the dog."

Oleg Gazenko, Soviet Animal Scientist, 1993

This time, much to Korolev's relief, the R-7 settled into its climb with good temper and the mission proceeded smoothly. Both dogs made it safely back to the ground after 17 orbits. There was great excitement in the world's press, and Soviet leader Nikita Khrushchev was delighted. Privately, Korolev and the space doctors were disturbed by a small incident during the flight. Belka became dizzy with the weightlessness and vomited into the cabin. Did this mean that humans would also become ill up there? Cameras in the ship recorded the dogs' behavior. Obviously the journey had not entertained them, although they seemed fine once they were back on the ground.

There was no longer any fear that humans might mysteriously disintegrate inside their capsules. On September 19, 1960, Korolev formally submitted his proposal for a manned flight, and the Central Committee of the Communist Party approved his request. Canine cosmonauts had proved that humans could survive a trip into orbit.

Oleg Gazenko, the animal scientist in charge of the dog flights, was proud that Strelka and Belka had

Ham the chimpanzee, above, inspects some of his flight hardware prior to his launch aboard a Mercury-Redstone combination opposite.

UNITED

"When you consider that the Mercury spacecraft, which is not much bigger than a standard telephone box, has thousands upon thousands of separate parts in it, and seven miles of wire winding back and forth, linking all the parts together, you get some idea of the problem of designing and building it and making it work."
Mercury astronaut John Glenn, 1962

returned safely to the earth to live comfortable lives as much-petted celebrities. Strelka even had a litter of six puppies, one of which, Pushinka, was sent to the children of America's new president, John F. Kennedy, as a diplomatic gift. Kennedy had the good grace to accept it, and Pushinka had a litter of pups with another White House dog, Charlie. But when it came to Laika's casual destruction, Russian space officials came to regret the haste with which that dog's mission had been prepared, under the rush of political imperative. Four decades later, Gazenko was free to admit, "Work with animals is a source of suffering to all of us. We treat them like babies who cannot speak. The more time passes, the more I'm sorry about it. We shouldn't have done it. We did not learn enough from the mission to justify the death of the dog."

The early American Mercury program focused on primates, including two chimpanzees, Ham and Enos. This gave NASA's critics an opportunity to mock, for it was easy to suggest that apes and humans were pretty much the same thing: man-shaped specimens to be "wired up the kazoo" with medical sensors, stuffed inside the little Mercury capsules like canned meat and fired from the circus master's cannon across the celestial ring. One newspaper cartoon had a pair of apes casually strolling away from a just-landed capsule. One turns to the other and says, "I think we're a little behind the Russians but slightly ahead of the Americans." Another cartoon showed an equally blasé simian pointing to a diagram on a wall, briefing his human colleagues on what to expect during a Mercury mission. "...Then, at 900,000 feet, you'll get the feeling that you must have a banana."

A Mercury without any primates on board was launched on February 21, 1961, for a successful test of the capsule's heat shield and parachute recovery systems. It was also a test of the powerful Atlas missile, modified to carry the little spacecraft into orbit. There had been a great deal of nervousness surrounding this flight, for if anything went wrong, it would call into question America's nuclear strike capability. Atlas's main role was as a weapons carrier, not a space launcher. Next day, NASA announced that Alan Shepard, John Glenn and Virgil Grissom had been short-listed as candidates to make the first manned flight. This trip was not to be a full orbit. No one yet knew how a human might stand up to a prolonged period of weightlessness, and the Atlas was not yet considered safe enough to carry one. One lucky astronaut would undergo a brief 15-minute suborbital trip aboard the smaller but reliable Redstone rocket built by von Braun and his Army Ballistic Missile Agency team, now working within NASA.

Technicians at the McDonnell Aircraft plant in St. Louis, Missouri, assemble a Mercury capsule under super-clean conditions. NASA demanded the most exacting standards of construction.

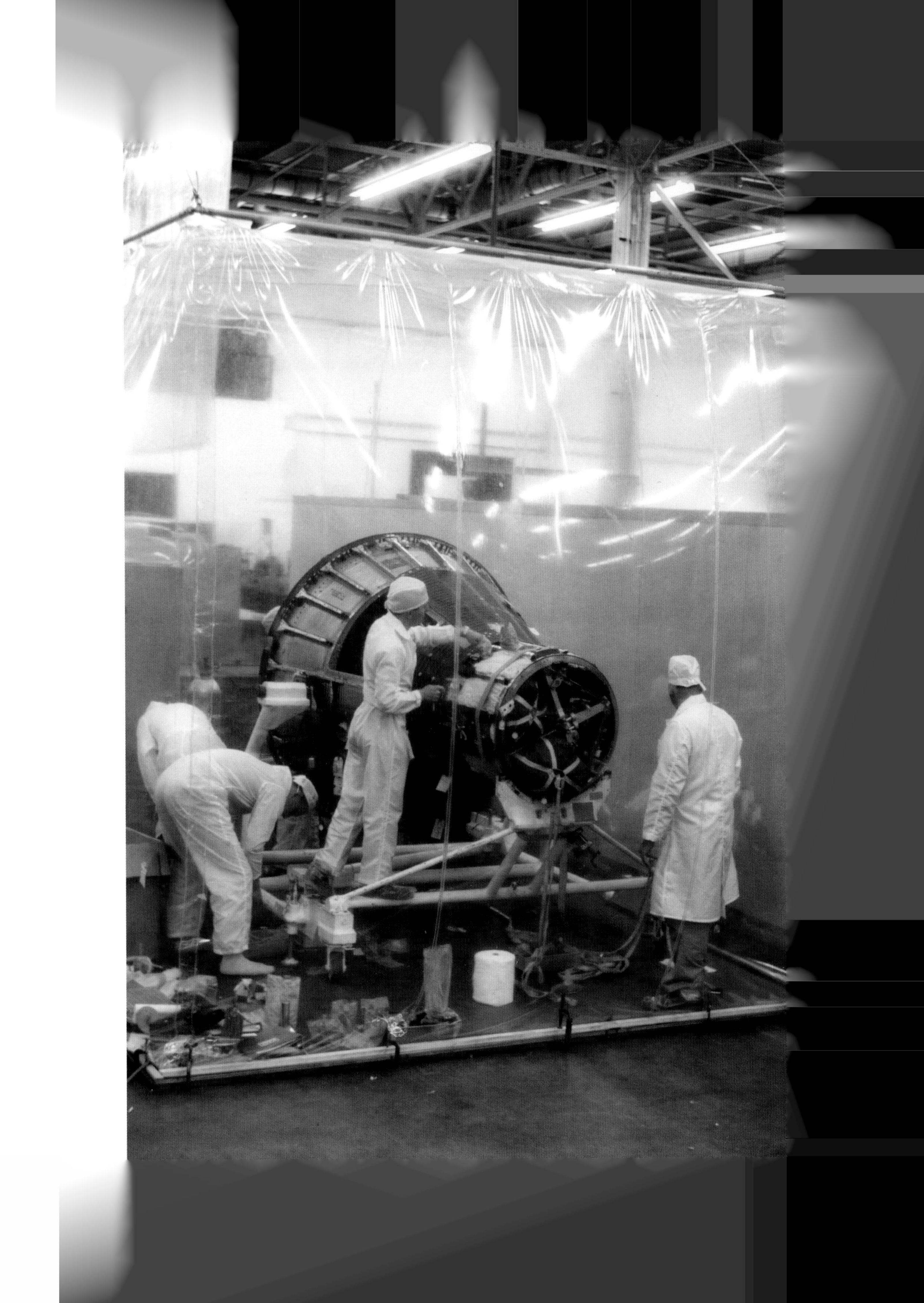

At von Braun's insistence, a last-minute snag had to be fixed before the mission could go ahead. Space historian John Logsdon takes up the story: "The first American manned flight should have happened in March 1961, but a previous Mercury test with the Redstone on January 31 had a chimpanzee called Ham on board. The retro rockets fired late, sending Ham a hundred miles downrange of the correct splashdown zone, and it took several hours to recover him—which made for one very unhappy chimpanzee by the way. The technical problem was simple, easy to fix, but they had to do another test of the Mercury before committing a human. So it's an interesting question: what would have happened if we had made that original date? I think history would have worked out very differently."

Certainly if a NASA astronaut had been first into space there would have been no particular impetus for President Kennedy to set his famous "before this decade is out" target for a lunar landing. It might well have been up to the Soviets to lay down that particular challenge. It is even conceivable, if Ham's trip had gone a little more smoothly, that we would still be awaiting the first manned mission to the moon today.

NASA's official log of the Mercury project records that Ham appeared to be in good condition after his ordeal. But some time later, "when he was shown the spacecraft, it was visually apparent that he had no further interest in cooperating with the spaceflight program." He certainly earned the long and comfortable retirement awaiting him in a special zoo.

Electrical short-circuits during his flight and water leakages after splashdown played havoc with the Mercury's cabin, causing the chimp considerable distress. No one wanted to take any chances with a capsule containing the first astronaut, and a seemingly minor delay entered the schedule. As a result NASA lost the race to launch the first man into space by just twenty-three days. And so the space race began.

Ham is succesfully recovered after an uncomfortable flight aboard his Mercury capsule.

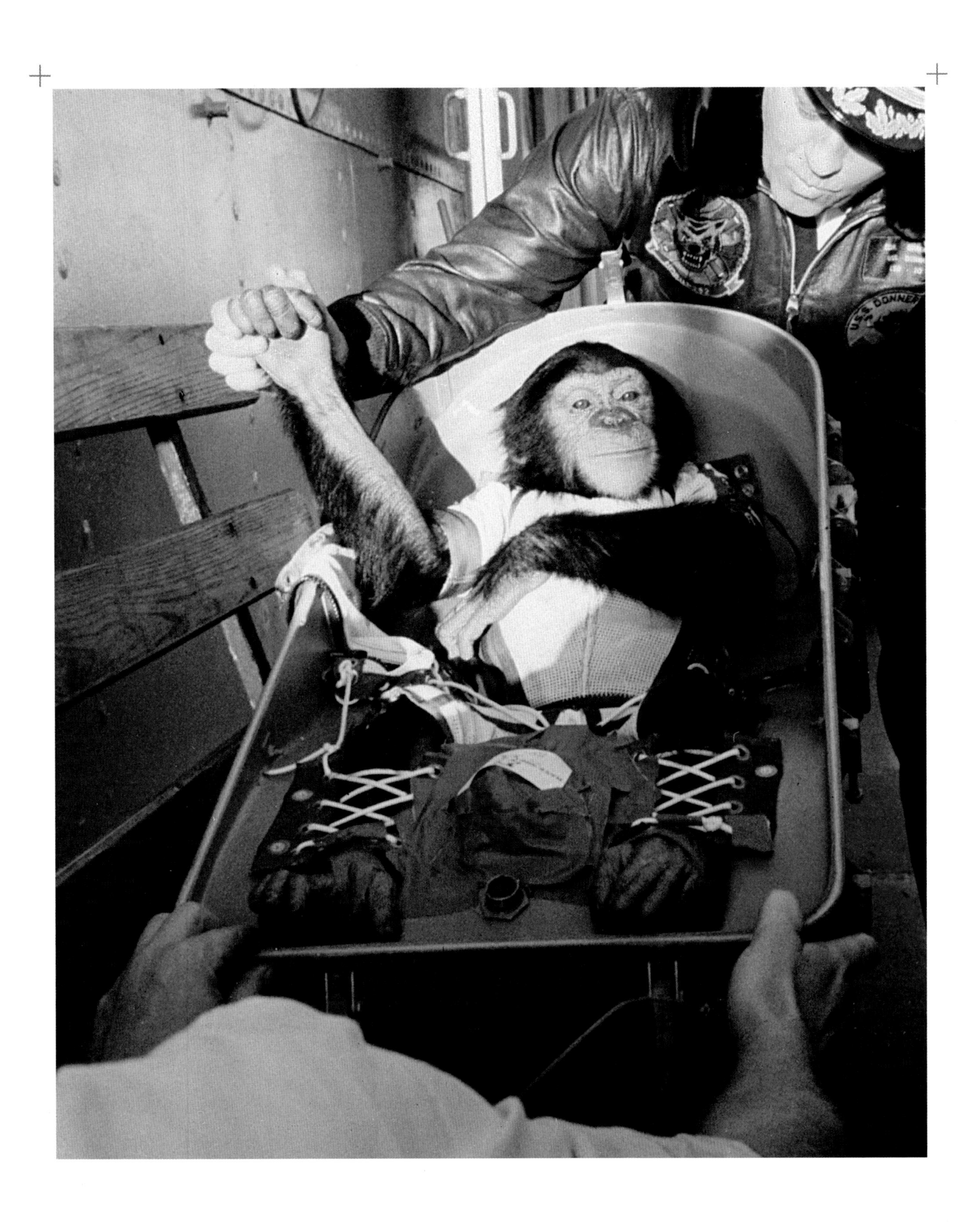
U.S.S. DONNER

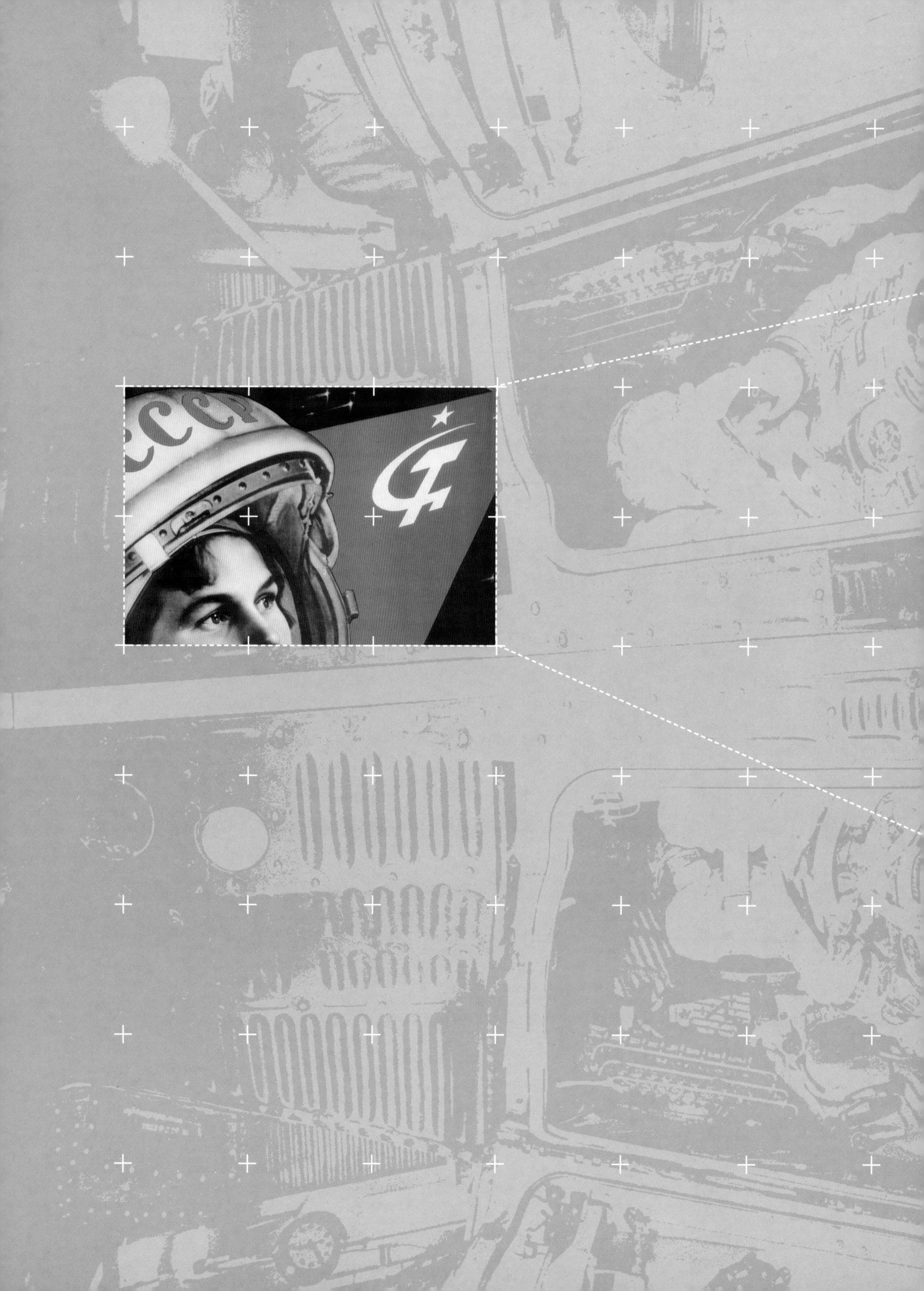

02 THE RACE

"The whole of my life seems to be condensed into this one wonderful moment. Everything that I have been, everything I have done, was for this. Of course I'm happy! To take part in new discoveries, to be the first to enter the cosmos, to engage in a single-handed duel with nature... Could anyone dream of more?"
Yuri Gagarin, 1961

THE RACE

By the spring of 1961, the CIA and other American intelligence agencies had been warning for some time that the Soviets were preparing to launch a manned capsule. The rumors rapidly solidified into fact. On April 11, 1961, President Kennedy appeared on an NBC early-evening television program sponsored by Crest toothpaste. He and his wife Jacqueline talked with reporters Sander Vanocur and Ray Scherer about the difficulties of raising their small children, and about the President's personal working style. Kennedy happened to remark that political events often appeared more complicated when viewed from inside the Oval Office than they did to the outside world. Even as he smiled for the television cameras, he knew from Intelligence reports that a serious embarrassment awaited him in just a few hours' time. There was little he could do except brace himself.

At 1:07 a.m. Eastern Standard Time, NATO radar stations recorded the launch of a Soviet R-7 rocket, and fifteen minutes later a radio monitoring post in the Aleutian Islands detected unmistakable signs of live dialogue with its human occupant. At 5:30 a.m. Washington time the Moscow News radio channel announced the latest Soviet triumph in space. An alert journalist called NASA's launch center at Cape Canaveral in Florida to ask, could America catch up? Press officer John "Shorty" Powers was trying to grab a few hours' rest in his cramped office cot. He and many other NASA staffers were working sixteen-hour days in the run-up to Mercury's first manned flight. When the phone at his side rang in the pre-dawn silence, he was irritable and unprepared. "Hey, what is this!" he yelled into the phone. "We're all asleep down here!" Later that morning the headlines read, SOVIETS PUT MAN IN SPACE. Spokesman says, U.S. ASLEEP.

Yuri Gagarin, a 27-year-old Soviet Air Force pilot, had just become the first human ever to fly in space, aboard a small capsule named Vostok (meaning east). Kennedy held an uncomfortable press conference. Normally a self-confident and eloquent public performer, he seemed distinctly less sure of himself than usual. He was asked, "Mr. President, a member of Congress today said he was tired of seeing the United States coming second to Russia in the space field. What is the prospect that we will catch up?"

He replied, "However tired anybody may be—and no one is more tired than I am—it is going to take some time. The news will be worse before it gets better. We are, I hope, going to go into other areas where we can be first, and which will bring perhaps more long-range benefits to mankind. But we are behind."

Gagarin's short journey into space was one of the most important events of the twentieth century—not just for Russia, but for America too, where an industrial shake-up of colossal proportions was unleashed in response. Much of the fabric of modern technology was designed with the space race (and a potential missile war) in mind. Microchips were developed because 1950s circuitry was too big, too heavy and too delicate to fit inside rockets and missiles. The Internet emerged from an attack-proof communications network laid down by ARPA, the Advanced Research Projects Agency, a pre-NASA government department that planned for America's future in space, among other things. The global communications industry developed with incredible speed after the invention of satellites. In all likelihood these technologies would have come along anyway, but probably not as fast as they did. And all because Yuri Gagarin, a good-natured young jet pilot and now the world's first space traveler, had thrown down a challenge to the most powerful nation on earth.

Dr. John Logsdon, who heads the Space Policy Institute in Washington D.C. and has advised a succession of presidents, explains the impact of Gagarin's flight on the American psyche. "It was a sudden rebalancing of our power relationship with the Soviet Union, because of the clear demonstration that,

Gagarin, above left, crowds in Moscow celebrate Gagarin's flight, above right, and a Vostok spacecraft opposite, on display in the mid-1960s.

if they wanted to, they could send a nuclear warhead across intercontinental distances, right into the heart of 'Fortress America'. There was an uproar: how did we get beaten by this supposedly backward country?"

Kennedy was agitated at the global response to Gagarin's flight. He paced his office at the White House, asking his advisors, "What can we do? How can we catch up?" His science advisor Jerome Wiesner opposed any reckless rush, but the President wanted an urgent response. "If somebody can just tell me how to catch up. Let's find somebody. Anybody. I don't care if it's the janitor over there, if he knows how." He deliberately made these remarks within earshot of Hugh Sidey, a senior journalist from *Life* magazine. All of a sudden the president wanted to be seen as an advocate for space. Three days later he suffered another more serious defeat. A 1,300-strong force of exiled Cubans supported by the CIA landed at the Bay of Pigs in Cuba with the aim of destroying Fidel Castro's communist regime. Kennedy had approved the scheme personally, but Castro's troops learned of the operation ahead of time and were waiting on the beaches. The raid was a disaster. The Kennedy administration seemed to be faltering in its first 100 days, the traditional honeymoon period during which a new president was supposed to stamp his particular vision on the country. He immediately turned to space as a means of reviving his credibility. In a pivotal memo of April 20, he asked Vice President Lyndon Johnson:

"Do we have a chance of beating the Soviets by putting a laboratory in space, or by a trip around the moon, or by a rocket to land on the moon, or by a rocket to go to the moon and back with a man? Is there any other space program which promises dramatic results in which we could win?"

"I don't know what you could say about a day in which you have seen four beautiful sunsets."
Mercury astronaut John Glenn, February 1962

Gordon Cooper prior to his two-day Mercury mission in May 1963. NASA's first astronauts were household names, idolized by millions of people as futuristic silver-suited heroes.

Johnson convened a panel of experts to help decide the matter, but he already knew which option he favored. Wiesner commented afterwards that, "Johnson went around the room saying, 'We've got a terribly important decision to make. Shall we put a man on the moon?' And everybody said, 'Yes.' And he said, 'Thank you,' and reported to the President that the panel said we should put a man on the moon." Or to put it another way, in the words of Pulitzer Prize-winning historian Walter McDougall, "Johnson sent the President a report so loaded with assumptions that a moon landing was the inescapable conclusion." As a powerful Democrat senator and staunch critic of President Eisenhower, Johnson had overseen the creation of NASA in 1958 as a response to Sputnik. Now he steered it towards the moon because he believed in the value of that goal, not just as a sturdy reply to Soviet domination, but also as a genuine expression of the Democrats' New Deal spirit. His politics had been forged in the Roosevelt era. Kennedy went along with the lunar project because it was expedient for him at the time, but he had less zeal for space than history tends to record.

A final decision hinged on NASA getting their manned space program off the ground. On May 5, just 23 days after Gagarin had flown, US astronaut Alan Shepard was launched atop a small Redstone booster. His flight wasn't an orbit, merely a ballistic hop of 15 minutes' duration. In contrast to Vostok's orbital velocity of 15,500 mph, Shepard's Mercury achieved only 5,000 mph. Vostok girdled the globe while the Mercury splashed down in the Atlantic just 320 miles from its launch site. But this cannonball flight was enough to prove NASA's basic capabilities.

Stunned by the potential costs, Kennedy nevertheless decided to support NASA's Apollo lunar project. In a historic speech before Congress on May 25, 1961, he said, "I believe that this nation should commit itself to achieving the goal, before this decade is out, of landing a man on the moon, and returning him safely to the earth. No single space project in this period will be more impressive to mankind, or more important for the long-range exploration of space, and none will be so difficult or expensive to accomplish."

On July 21, 1961, astronaut Virgil "Gus" Grissom flew another suborbital curve in a Mercury capsule atop a Redstone ballistic missile. His mission nearly ended in disaster when the hatch on the capsule blew off shortly after he splashed down. He clambered out of the waterlogged craft without his helmet, and water poured through his neck ring and into his spacesuit. He tried to signal the approaching rescue helicopters for help but was amazed to see them fly over the capsule instead of coming directly to his aid. The pilots thought Grissom was waving, not drowning. Eventually Grissom was rescued, but his capsule sank to the bottom of the Atlantic beyond all hope of recovery.

NASA downplayed the fact that their astronaut was nearly lost at sea and hailed Grissom's mission as a near-perfect success. But the moon seemed a long way off. No one was even sure if NASA could really get there. Certainly it couldn't in a waterlogged Mercury.

L. G. COOPER

Pressure from Russia

Alan Shepard, America's first man in space, in his cramped Mercury capsule just before launch on May 5, 1961.

> Soviet leader Nikita Khrushchev and the Politburo had no particular interest in flying to the moon. Instead Khrushchev pressured Korolev for short-term rocket victories in the hope of demoralizing the American space effort before it became unstoppable. As long as Korolev continued to deliver results with the Vostok and his converted R-7 missile, he could count on the Kremlin's continuing support.

On August 6, Gherman Titov took off from Baikonur and flew 17 orbits aboard the second manned Vostok, staying aloft for 24 hours and flying the craft manually for a short period. Just like the dog Belka, Titov was extremely nauseous during his flight. The heating in the cabin broke down so that he nearly froze, and his retro rocket pack did not separate cleanly before reentry, which gave him some cause for concern.

Vostok cosmonauts had to eject from their craft in the high atmosphere and parachute to the ground, for they could not have survived the capsule's meteor-like touchdown on the Russian steppes. However, Titov's ejection and landing in the Saratov region was not without its hazards. "Under my parachute I passed about 150 feet from a railway line, and I thought I was going to hit the train that was passing. Then, about 15 feet from the ground, a gust of wind turned me around so that I was moving backwards when I hit the ground, and I rolled over three times. The wind was very brisk and it caught the parachute again, so I was dragged along the ground for some distance. When I opened my helmet the rim of the faceplate was scooping up soil. You know, the farmers in Saratov had done their ploughing quite well that season, otherwise my landing would have been even harder."

Sick, exhausted, bruised but alive, Titov was the first man to spend an entire day in space, and the first to make multiple orbits around the earth. Perhaps he took additional satisfaction from beating Gagarin at his own game. Saratov is about 900 miles from the Baikonur launch complex, which means that Gagarin's historic but incomplete "first orbit" on April 12 fell short by that distance. So Titov was actually the first man to complete a full circle.

One week after Titov's landing, construction of the Berlin Wall began. According to Korolev's biographer James Harford, Khrushchev ordered the timing of Titov's flight deliberately to foster the German Democratic Republic's loyalty to Moscow. However, there was more to the relationship between Khrushchev and Kennedy than simple-minded East-West aggression. It is becoming clear that the two men considered the possibility of a collaboration in space—perhaps a nominal face-saving one that would allow both sides to step down from the brink of a colossally expensive and impractical adventure. Space historian John Logsdon says, "Kennedy was ambivalent about space even after he'd announced Apollo. He sent feelers through his brother Bobby into back channels at the Kremlin."

But by now, it was too late for either side to turn away from the contest. The momentum was unstoppable.

"Those guys were being recruited to become passengers in capsules carried into space by rockets which regularly blew up."
Apollo 15 commander Dave Scott, 2004

Above, Gordon Cooper is helped into his Mercury capsule by the launch pad team. An Atlas rocket is made ready opposite, to hurl Cooper's Mercury into orbit.

UNITED STATES
130D

"Well, the Lord giveth, and the Lord taketh away. The mantle of Cold Warrior of the heavens had been placed on their shoulders without their asking for it; and now it would be taken away again without their knowing that either. It would have been impossible for them to realize that the day might come when Americans would hear their names and say, 'Oh yes – now which one was he?'"
Tom Wolfe, *The Right Stuff*, 1979

"For several minutes Gagarin's reentry ball and its rear module remained tied together like a pair of boots with the laces knotted. The nightmarish ensemble tumbled end over end in its headlong rush to earth."
Piers Bizony & Jamie Doran, *Starman*, 1998

Opposite, a recently landed Vostok capsule lies in a field like some strange meteorite. The Soviet authorities concealed the fact that the first cosmonauts ejected from their capsules just before touchdown and descended to earth under separate parachutes.

A celebratory poster showing Valentina Tereshkova in her space helmet.

The Forgotten Women of Space

> After February 20, 1962, when John Glenn was launched on the more powerful Atlas booster and completed three full orbits of the earth, NASA felt better prepared to take on the gigantic challenges ahead. Glenn was applauded by U.S. citizens just as Gagarin had been in Soviet Russia.

Russia's Vostok series continued, while Khrushchev pressed for yet more rocket "firsts". As far as can be judged, it was his idea to launch the kind of person that neither side had yet sent into space: a woman.

By 1962, up to 400 Russian women had been screened for a possible flight. It just so happened that parachuting was encouraged as a hobby for women at that time. This was an invaluable skill, given the Vostok capsule's method for crew escape and landing. On February 16, five champion parachutists were selected for training: Valentina Ponomaryova, Tatyana Kuznestova, Irina Solovyova, Valentina Tereshkova and Zhanna Yorkina. The favored candidate to emerge was Tereshkova, a 25-year-old textile worker with 58 parachute jumps to her credit. She was ideal for Khrushchev's purposes: fit, good-looking, smart enough for the challenges of space training, but not so advanced in her education that she couldn't represent the ordinary peasant and working classes.

Gherman Titov acknowledged the sexism with which Tereshkova and her colleagues were greeted when they first arrived at the Star City cosmonaut training center near Moscow. "We didn't believe that women belonged anywhere near a flying machine. At that time, we thought that only men could carry out all the tasks involved in spaceflight. But in the end we saw it was right to have female cosmonauts, and we soon thought of them as good guys, just like us."

Tereshkova's eventual flight on June 16, 1963, aboard Vostok VI, presented very few fresh technical challenges. She made a close approach to Vostok V, piloted by cosmonaut Valeri Bykovsky; but the real trick was in the timing of the respective launches. Her success gave Khrushchev an opportunity to gloat about "the equality of men and women in our country". Once this propagandist venture had been successfully completed, the women's cosmonaut squad was quietly disbanded. On November 3, 1963, Tereshkova and Vostok III cosmonaut Andrian Nikolayev were married in Moscow in a very public ceremony. It was the social event of the season, much enjoyed by Khrushchev. The marriage did not last.

Almost no one at the time took the trouble to remind the world that brave Russian women aviators had already proved themselves in combat during World War II. Stalin's terror purges had left the Red Army dangerously unprepared for the massive invasion of German troops in June 1941. As for air power, in some battle zones the only airplanes available were canvas-covered Polikarpov biplanes built in the 1920s. There were very few bombs or weapons for them to deploy, and no spare male pilots to fly them. In October 1941 Soviet women pilots were organized into combat regiments by Marina Raskova, a famous Russian long-

"Nothing means more to me than flying into space. Nothing in life is as important. I would do anything for the chance. This is something I would give my life for."
Jerrie Cobb, 1998

distance aviator. Her exploits had been approved in peacetime as a riposte to America's Amelia Earhart and Britain's Amy Johnson. Now she assembled a squad of female pilots to fly the ragged Polikarpovs.

Despite being vulnerable, under-equipped and underpowered, these planes and their pilots were soon dreaded by the German ground troops. They called them Nachthexen, "Night Witches". And little wonder. The planes ventured out in the pre-dawn darkness and dropped lengths of old railway track onto the German encampments while most of the troops were still asleep. There were no explosions, but the tracks bounced and smashed their way through tents and vehicles, wreaking havoc. They made no noise as they fell. It was utterly terrifying.

As the war progressed and Russia regained its military effectiveness, the Night Witches began dropping real bombs. Yak-1 fighter planes became available, and a fighter squadron commanded by Tamara Kazarinova flew over four thousand missions, dogfighting with German bomber escort planes.

Meanwhile in the United States more than a thousand Women's Airforce Service Pilots (WASPs) were ferrying fighters and bombers direct from the manufacturing plants to battle theaters around the world, under the leadership of record-breaking pilot Jacqueline Cochran. WASPs were seldom permitted to carry ammunition and had to take evasive action as best they could if they came under attack in the air. They worked tirelessly on long-haul flights, delivering vital equipment: cargo planes, bombers and high-performance fighters. Abruptly, at the end of the war, the WASPs were instructed to go back to their families. Many of them were shocked to find that the Air Force had not even bothered to buy them their train tickets home. It took half a century even before a proper national memorial to fallen WASP pilots was established.

Nation states rebuilding after the war tended to focus on home and hearth. Perhaps as a result, many women's personal ambitions were thwarted unless they were connected with the raising of families. But some exceptionally determined women did make careers in the aviation business. When NASA began recruiting astronauts for its early space programs, the air speed and altitude champion Jerrie Cobb decided to apply. Dr. Randy Lovelace, a NASA life sciences expert closely involved with the medical testing of astronauts, subjected Cobb to a blistering series of tests at his private clinic—without official permission. She scored so well that Lovelace could see no reason to ban her from spaceflight. Cobb then recruited more women to take the tests, with funding from Jacqueline Cochran. Twelve more women were chosen for possible astronaut training: Bernice Steadman, Janey Hart, Jerri Truhill, Rhea Woltman, Sarah Ratley, Jan and Marion Dietrich, Myrtle Cagle, Irene Leverton, Gene Nora Jessen, Jean Hixson and Wally Funk.

But it came to nothing. Although NASA appointed Cobb as a Special Advisor in 1961, she never flew in space, and neither did her team mates. NASA's official

Jerrie Cobb tries out the controls of a spacecraft simulator that tumbles in three axes. An expert pilot, she passed all the tests with flying colors, but never flew into orbit.

position was that only pilots with military experience could join the astronaut corp. Since American women were not at that time allowed to fly military missions, they were disqualified from the space adventure.

The simple truth is that America was not ready to see a female astronaut blown to pieces in a launchpad disaster or slowly suffocating in a doomed orbit. Risking life and limb was the preserve of men, while women's proper role was in the creation of life—at least, that's what the men thought. Bitterly disappointed, Cobb left NASA for the jungles of the Amazon, where she spent the next four decades as a solo pilot delivering food, medicine and other aid to the indigenous people. In 1981 she was nominated for a Nobel Peace Prize for this work.

In 1998, aged 67, Cobb was younger than 77-year-old John Glenn, the senator and veteran Mercury astronaut who flew aboard the space shuttle Discovery in October that year and was welcomed back to NASA as a returning hero. Glenn wanted to prove that age was no bar to flying in space. He also conducted useful medical research into some of the effects of ageing, including loss of bone density. (The biochemistry of bones in weightlessness is intimately connected with osteoporosis, a disease that afflicts many people in later life.) Cobb, however, was unable to follow Glenn's example. "Nothing means more to me than flying into space," she said at the time. "Nothing in life is as important. I would do anything for the chance. This is something I would give my life for."

In some senses, she already has. Her example as a pilot and her early work at NASA helped pave the way for today's female astronauts, even if their chance to fly was a long time coming. Sally Ride became the first American woman in space in 1983, flying aboard the shuttle Challenger. Three years later she served on the panel that investigated the loss of that same ship.

Women are at the forefront of the space adventure today. In July 1999 the STS-93 Columbia mission was the first to be commanded by a woman, Eileen Collins. In July 2005 she commanded Discovery's trip to the Space Station in a tense flight following the loss of a second shuttle, Columbia, in 2003.

Women of space: left, NASA's Kathy Sullivan, right, Claudie Haigneré from the European Space Agency ESA, and Sally Ride, opposite, America's first woman in space.

RIGHT AUDIO
VOLUME
ICOM
TACAN

A Walk in Space, then the Wolves

> As NASA announced plans to create a new two-man capsule, the Gemini, in the Soviet Union Korolev took a risk. He decided to adapt the Vostok to carry two cosmonauts in the ball-shaped crew module, and even three if they gave up their spacesuits. This new seating arrangement was purely cosmetic; it didn't make the Vostok any better, just more cramped and significantly more dangerous. The bulky ejection seats had to be sacrificed in order to make room for the extra crew, so the landing (albeit cushioned by newly installed retro rockets) would be a tough one. And there was no chance of escape if anything went wrong on the launchpad. This new craft was called Voskhod (Sunrise).

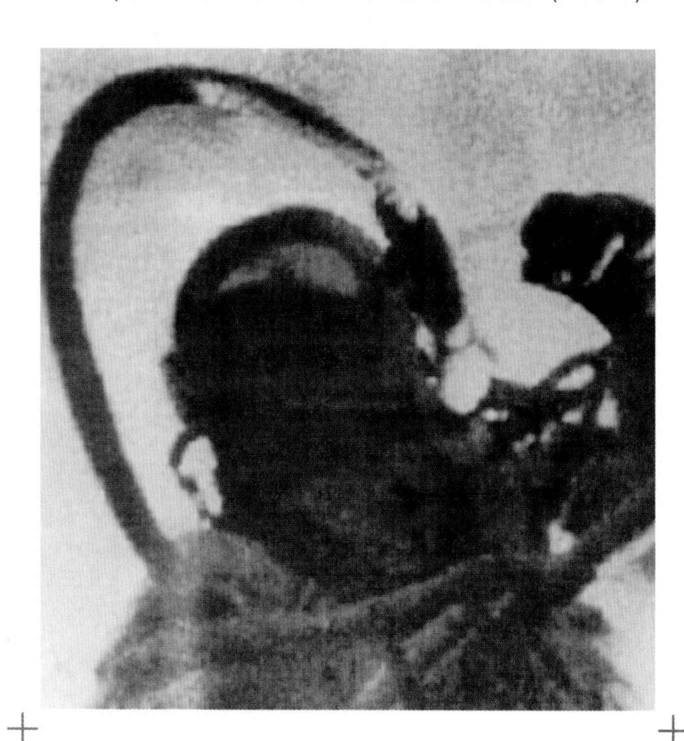

On October 12, 1964, Korolev made good on his promises to Khrushchev by successfully launching Voskhod I with three crew aboard: Vladimir Komarov, Konstantin Feoktistov—Voskhod's senior engineer trusting his life to his own design—and Boris Yegorov. The mission flew too late to benefit Khrushchev. The capsule came home on October 13, and on the very next day he was deposed and replaced by Leonid Brezhnev. A letter from the Politburo informed him that he had just resigned due to age and increasing ill-health.

The Voskhod II mission took off on March 18, 1965, ahead of NASA's first two-man Gemini mission by six days. This time there were only two crewmen in the cabin, to make room for their bulky spacesuits. Pavel Belyayev remained inside while his co-pilot Alexei Leonov squeezed into the flexible airlock and pushed himself out of the capsule. For ten minutes he enjoyed the exhilarating sensation of spacewalking, and then began to pull himself back into the ship—only to discover that his suit, at full pressure, had ballooned outwards so that he could no longer fit into the airlock. Dangerously exhausted by his efforts, Leonov had to let some of the air out of his suit to collapse it so that he could squeeze back into the spacecraft.

Left and opposite, from the days before color TV, transmitting any kind of an image from space seemed miraculous. Here, Alexei Leonov's ghostly form drifts in a hazy void.

"Nothing will ever compare to the exhilaration I felt in that moment. No matter how much time has passed, I can still remember quite clearly my emotions."

First spacewalker,
Alexei Leonov, 2005

Before returning home the next day, Belyayev realized that the ship was not correctly lined up for the braking rocket burn, so he shut down the automatic guidance systems before they could make matters any worse. With help from Korolev and ground control, he and Leonov ignited the braking motors manually on the next orbit, displacing their originally intended landing coordinates by 1,200 miles. The capsule descended onto snow-covered Siberian wilderness. It smashed into a dense cluster of fir trees and was wedged about 10 feet above the ground between two sturdy trunks.

Meanwhile the recovery teams were 1,200 miles away in the zone where they had originally expected the capsule to come down before its navigation systems failed. The cosmonauts had to spend a restless, frozen night waiting to be picked up. They pushed open the capsule's hatch but did not dare climb down from their precarious perch because a pack of wolves was howling somewhere very nearby in the darkness.

These difficulties were not mentioned in the Soviet press reports, and the mission was a great propaganda triumph all around the world. Korolev's team received a stern phone call from the Kremlin, ordering that not a single word about the landing in Siberia should appear in the media.

Subterfuge aside, the fact remains that Leonov walked in space well ahead of his American rivals. Once again NASA had been trumped. Leonov's spacewalking rival Ed White did not get his chance until the second Gemini mission on June 3, 1965, nearly three months after Voskhod II.

A keen artist, Leonov set about designing a commemorative postage stamp showing his spacecraft. His designs had to be vetted by anxious KGB propaganda experts. "Everything was so secret. I drew a completely different spacecraft that wasn't anything like what we had really flown. And then they were satisfied."

It is one of the unfortunate aspects of space history that so few iconic high-quality images of early Russian space achievements have entered the public domain. Perhaps that excessive Soviet secrecy was their greatest mistake.

Above, Leonov's heavily disguised artwork depicts a Soviet capsule.
Opposite, an officially released photo of his spacewalk has been retouched in color so that it almost becomes a painted artwork rather than a photo.

СССР

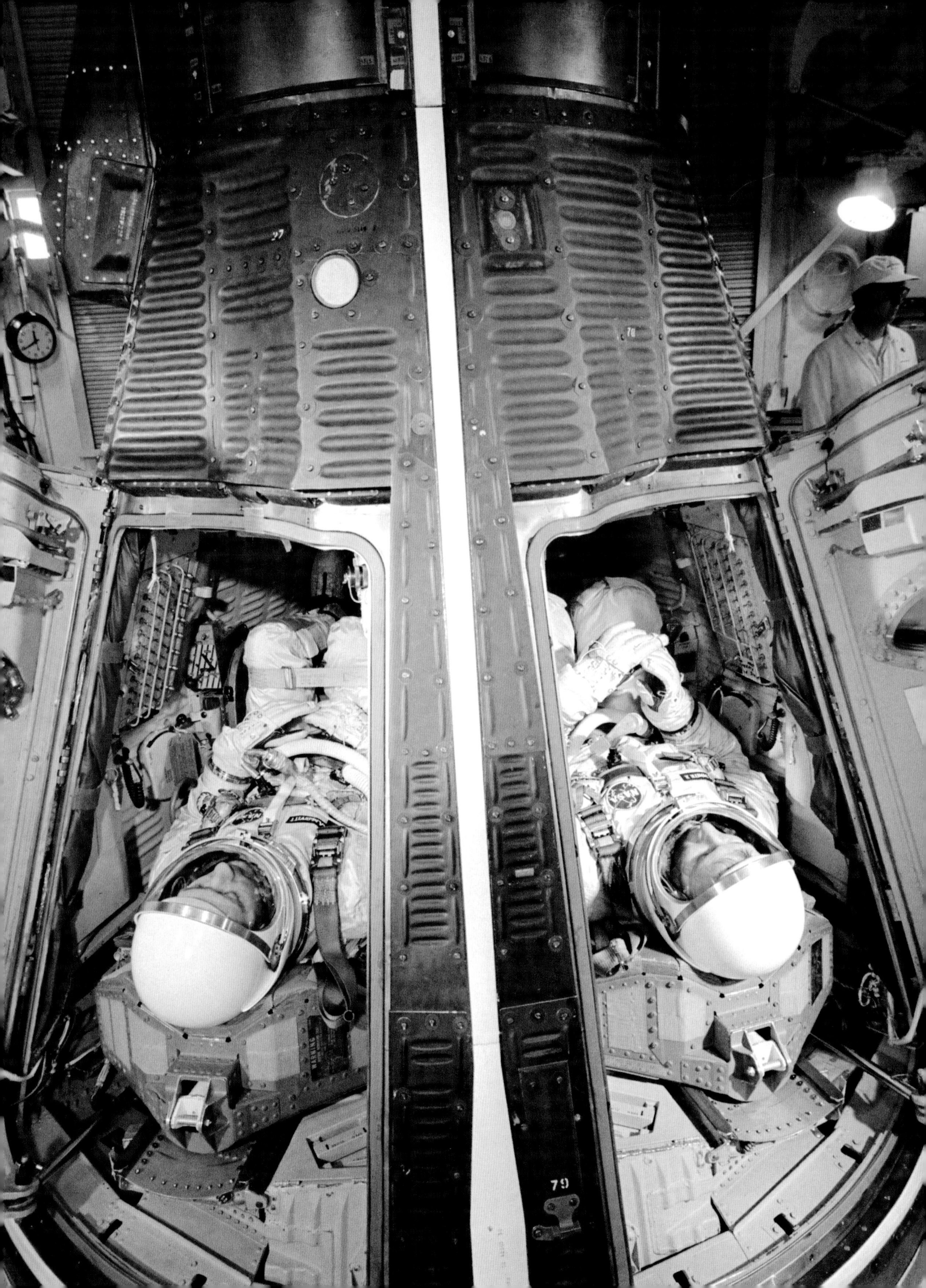

Gemini Rising

> The earliest Soviet capsules were operated mainly from the ground rather than by the pilots inside them. Space chiefs fretted that a cosmonaut might go crazy in orbit, overcome by the spiritual and psychological separation from his companions on earth, and the security services worried that he might defect to the West, deliberately reentering the atmosphere over foreign territory. The design of the instrument panels inside the Vostok capsules was not intended to give a pilot any authority over his own vehicle, but to take that control away from him. In an emergency he might be allowed to operate the controls for a while, but only if he could prove his sanity first. Engineers devised a six-digit keypad that would unlock the navigation systems and let the pilot steer his own ship if manual control became necessary. He'd be told the keypad combination only if mission directors on the ground decided he was mentally fit for the job.

Korolev fought against this paranoid security measure. Why would a pilot be given control of the ship? Presumably because the automatic systems had failed and he needed to take over. But if the ship started to tumble out of control, the radio link with earth might be interrupted just at the point when the pilot really needed to hear the secret code that would release his manual controls. The keypad idea seemed more dangerous than just leaving things be.

In NASA's early years, a similar controversy raged over the extent to which astronauts should be able to steer their own ships in the alien environment of space. As

"We have a saying in Russia: If your altitude is higher than zero, and your speed is faster than zero, then you have a risk. We have pretty high speed and a pretty high altitude, but that's what we train for. That's our profession."
Cosmonaut Sergei Krikalev

Opposite, astronauts James McDivitt and Ed White are about to be sealed inside their Gemini spacecraft for a simulated launch in June 1965.

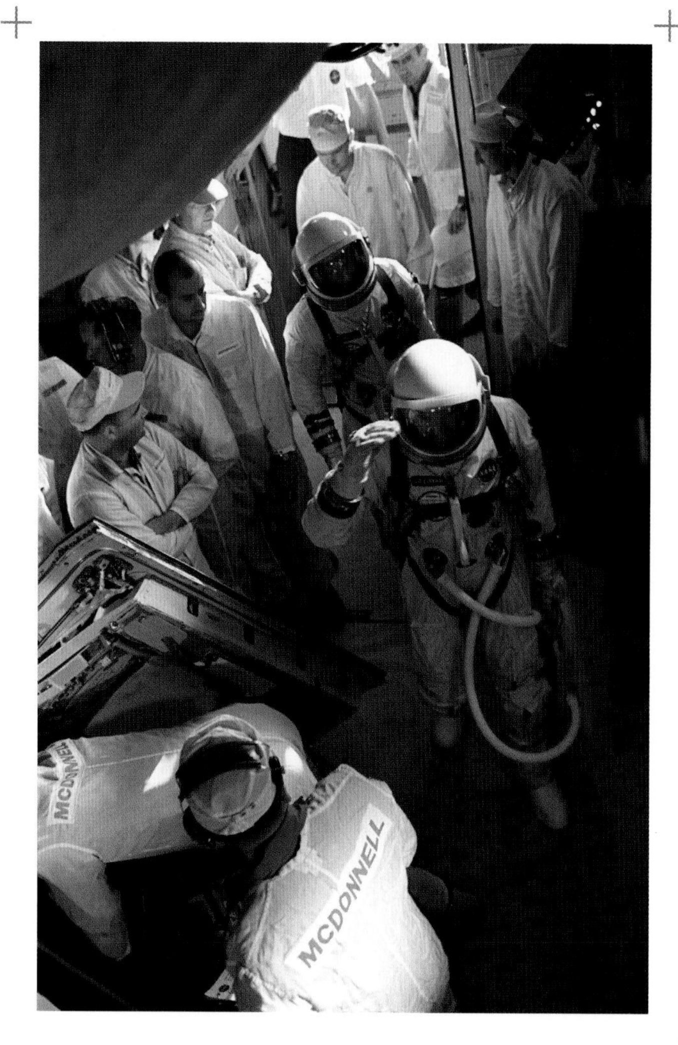

Left, Gordon Cooper salutes pad technicians as he and Charles Conrad prepare to take their seats aboard Gemini V in August 1965, and Conrad, opposite, seen through the window of the capsule.

Apollo loomed on the horizon, senior managers gradually realized that automatic control wouldn't be enough to guarantee successful missions. Gemini, the two-seat capsules that flew after Mercury and before Apollo in the mid-1960s, would test the astronauts' abilities as participants, rather than hapless passengers, in the space adventure.

Gemini crews took advantage of their high-profile appearances in *Life* magazine and on television to lobby for command over their own flights, or at least to obtain an equal partnership with their mission managers on the ground. It does not get much of the glory in space history today, but the two-seater Gemini craft was probably NASA's sportiest, with its huge gull-wing doors for space walking (or escape by ejection) and its complex array of thrusters. It was also the first space vehicle to carry an onboard computer to calculate its rocket firings.

Not that the computer was up to much by modern standards. Gemini veteran John Young gave an unflattering description. "Imagine a box with a lot of coat hangers bent in all directions. It was basically wiring connecting thousands of individual transistors, with tiny electromagnets for memory nodes, very crude compared to today's microprocessors." Yet this basic computer, with barely 15 kilobytes of memory, enabled Gemini to change orbits and accurately locate other target vehicles for docking: the most essential of all space maneuvers. Construction of the International Space Station relies on docking, but from 1965 to 1966 the priority was different. NASA's first astronauts needed to practice docking in anticipation of the moon missions while the Apollo lunar craft was still being designed. Gemini was their teaching tool.

While the public knew the two-man ship as Gemini, NASA insiders called it the "Gusmobile" in honor of Virgil "Gus" Grissom, the first astronaut to command one. Young, who flew the Gemini alongside Grissom, recalls that "Gus had a big hand in the design, and the way the cockpit was laid out. That ship was really his baby." Grissom spent many months at the McDonnell Douglas factory in St. Louis, helping to develop the ship for the astronauts' benefit rather than for the technicians on the ground. Or should that be, for his benefit? "At five-foot six, Gus was a couple of inches shorter than most of the rest of us," Young says. "The Gusmobile was a tight fit for the other guys, but it was worth it. Gus eliminated the problems we'd had on the one-man Mercury capsules before, where we were just like circus performers shot out of a cannon.

NASA

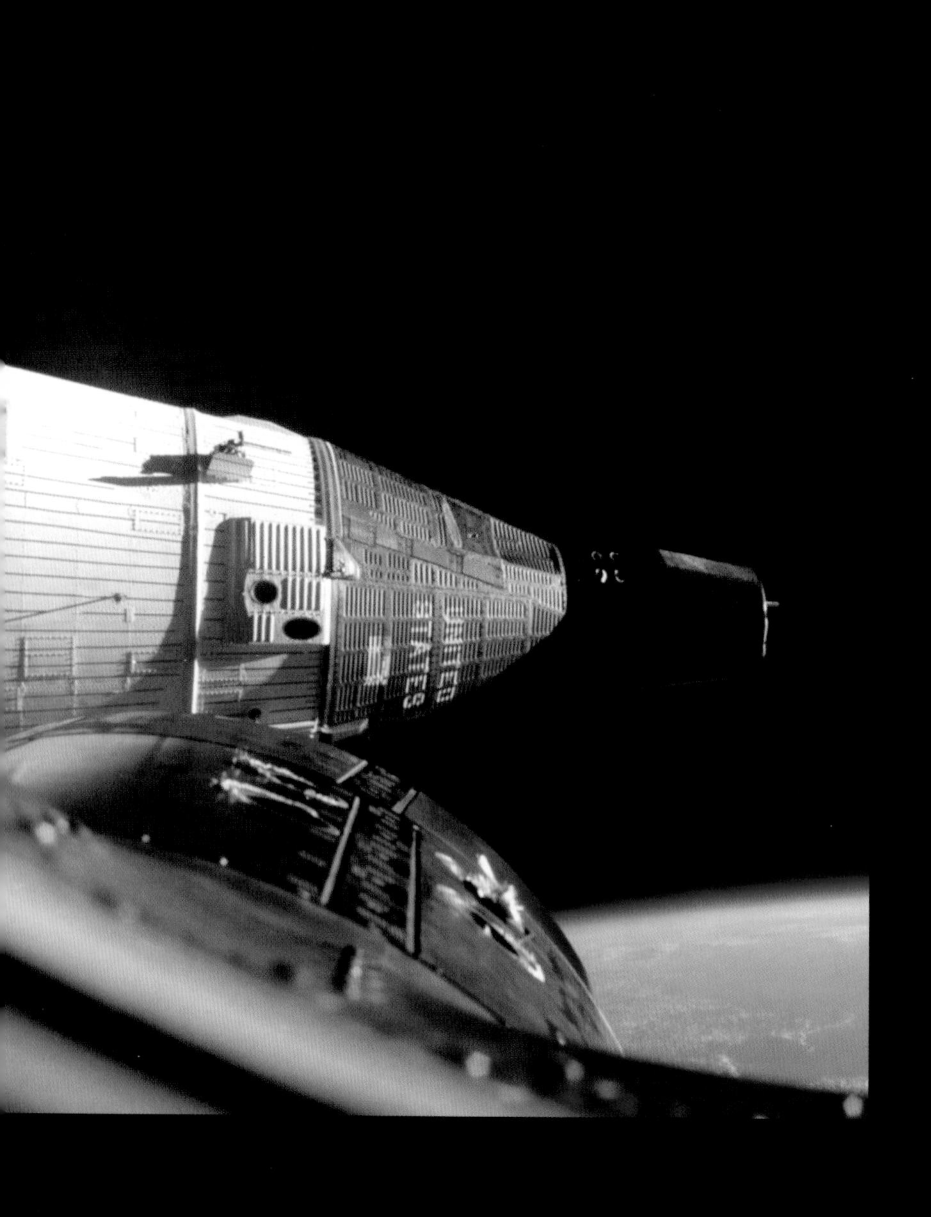

Opposite, Ed White makes his historic spacewalk from Gemini IV on June 3, 1965, and above, the first true space rendezvous takes place between Gemini VI and Gemini VII on December 15 of that same year.

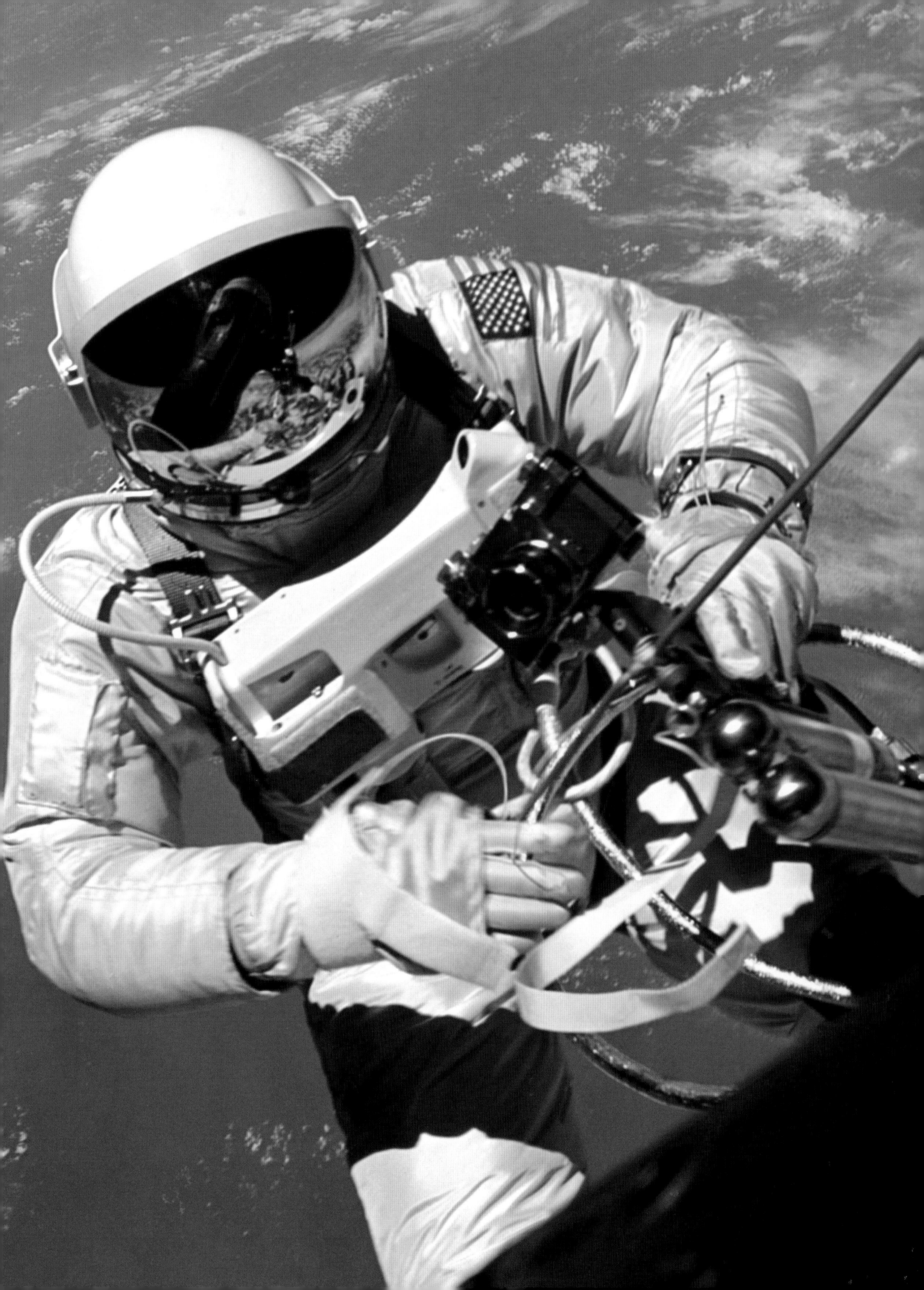

"No way did the Russians rendezvous. That was just a passing glance, like seeing a pretty girl on the pavement from your car window before the flow of traffic whizzes you on. With Gemini we can cut across the traffic and say hello. Now *that's* what you call a rendezvous."

Gemini VI Astronaut Wally Schirra, 1965

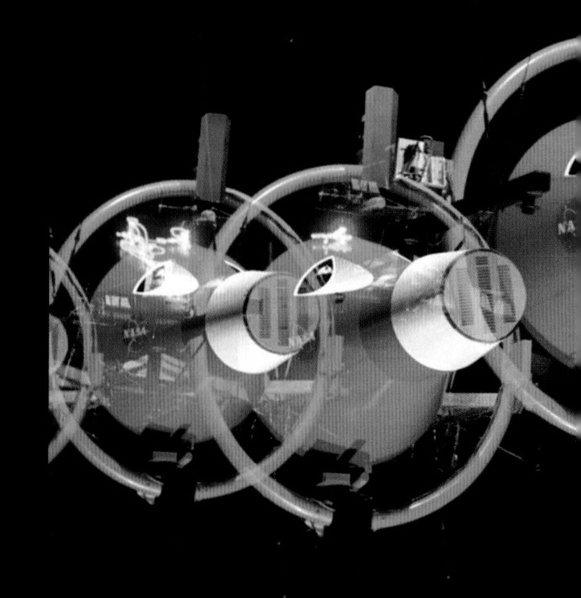

A mock-up Gemini capsule suspended on cables nudges towards a rendezvous with an Agena target vehicle in this multiple-exposure photograph from 1964.

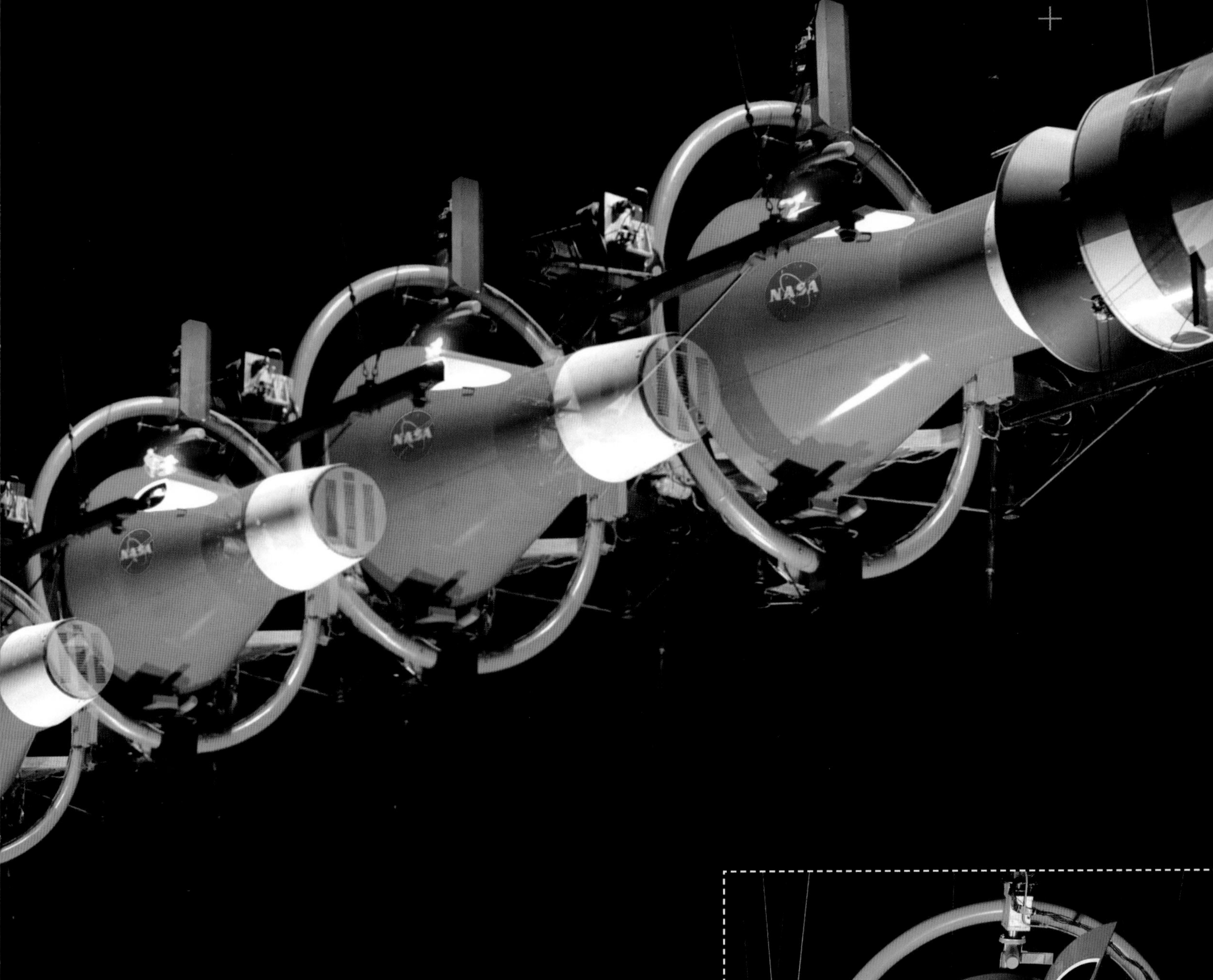
NASA
NASA
NASA

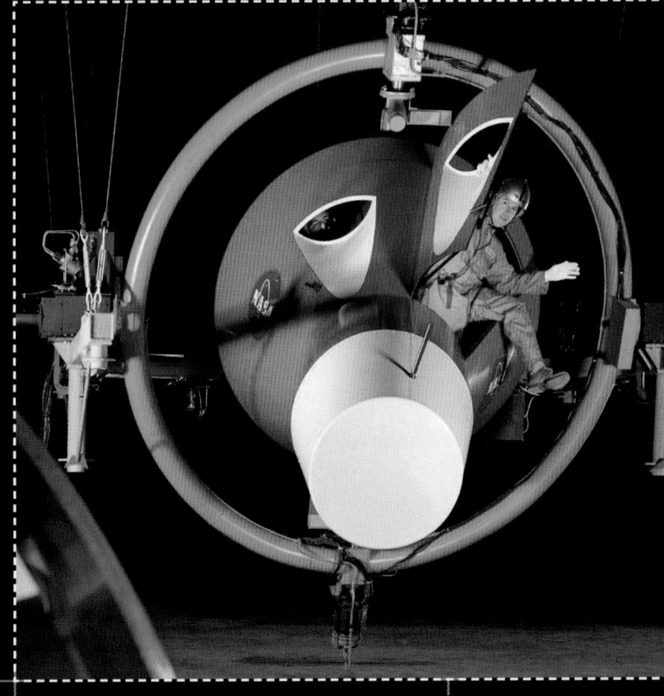

Splashdowns for NASA capsules could be hazardous and uncomfortable, below, but after Neil Armstrong and Dave Scott had run into trouble in space aboard Gemini VIII, they were relieved to make it back home, opposite.

"It would be a bitter irony if we ended our exciting voyage in space by getting lost at sea. But it could happen."
Astronaut Alan Shepard, 1962

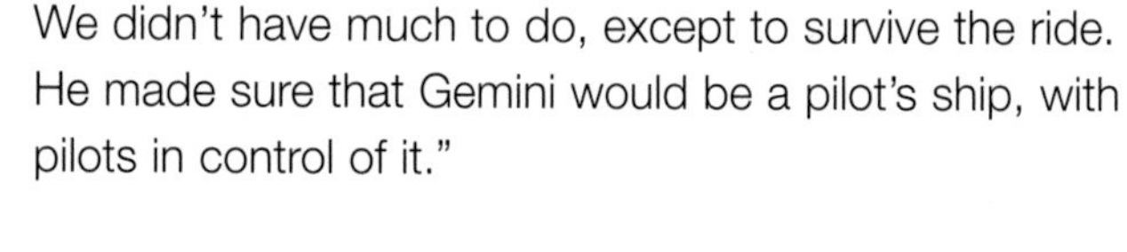

We didn't have much to do, except to survive the ride. He made sure that Gemini would be a pilot's ship, with pilots in control of it."

After two successful unmanned shots, Gemini III took off on March 23, 1965, with Grissom and Young at the controls for a quick five-hour orbital test of their Gusmobile. As it turned out, the machines performed better than the humans on this first manned trip of the capsule. Mission Control did not yet trust Gemini's new computer. Grissom had been instructed not to use it for reentry if the onboard displays disagreed with NASA's earlier predictions of where the capsule would splash down. The computer said Mission Control's figures were seriously wrong, so Grissom switched it off—and splashed down in the Atlantic more than 60 miles from the nearest rescue ships. Young said afterwards: "We knew a few days later, when NASA checked the figures, that if we'd gone with the computer we could have parked our capsule right alongside the recovery ship."

In June 1965, Jim McDivitt and Ed White spent their four-day Gemini mission without electronic support, McDivitt having to make his reentry calculations with help from the ground. Meanwhile the world's attention was focused on the first US space walk, conducted by White. He floated outside the capsule for half an hour, connected only by a thin umbilical cord. "This is great! I don't want to come back inside!" he said.

Subsequent Gemini missions demonstrated NASA's ability to rendezvous in space, and even dock. But

COLLAR
157-0

Glamorous activities in orbit overshadowed equally dramatic adventures in the earth's atmosphere during the development of exotic new machines for space. Bruce Peterson, left, prepares to climb aboard his M2-F2 "lifting body" prototype on May 10, 1967, prior to his release at high altitude from a converted B-52 aircraft opposite.

was this new skill a space "first" or had Russia already beaten NASA to it some three years earlier?

In August 1962, cosmonaut Andrian Nikolayev had launched in the same kind of Vostok capsule that propelled Yuri Gagarin into the history books. The next day Pavel Popovich went up in another Vostok. For the first time, two people were in space simultaneously, in two different ships. The Russians timed the launches so that the second Vostok would briefly come within 4 miles of the first: a cosmic clay pigeon shoot that enabled the Kremlin spin doctors to claim what looked like a "space rendezvous".

The two craft quickly drifted apart and could never have regained their initial close formation. However, appearances counted for a great deal. A number of space professionals in the West were fooled into thinking that the Soviets had developed genuine rendezvous skills. But as Gemini astronaut Walter Schirra commented to journalists, "No way did the Russians rendezvous. That was just a passing glance, like seeing a pretty girl on the pavement from your car window before the flow of traffic whizzes you on. With Gemini we can cut across the traffic and say hello. Now *that's* what you call a rendezvous."

Two astronauts made a special impression during the Gemini program. Buzz Aldrin was known as "Dr. Rendezvous" because of his detailed understanding of the physics of orbital docking, and Neil Armstrong showed his unflappable nature in what might so easily have been a disaster.

Armstrong and his copilot Dave Scott blasted off aboard Gemini VIII on March 16, 1966, to dock with an

unmanned Agena target vehicle launched earlier that same day. Armstrong steered the ship to the first ever space docking. This time there was no doubt about the "first". The Russians were years away from being able to hard-dock.

But it was a double-edged victory. CapCom Bill Anders strolled into Mission Control just as the docking was under way and colleague Jim Lovell handed the microphone to him, saying, "It's all pretty boring so far." Moments later, Armstrong's voice came on the line. "We've got serious problems. We're tumbling end over end." Armstrong immediately backed his capsule out of the Agena, only to find that the tumbling became worse. A thruster was jammed in the "on" position and Gemini VIII began to spin at close to ninety revolutions per minute. Armstrong and Scott could barely keep conscious. Armstrong took a bold decision and cut power to all the docking and control thrusters, hoping to kill the rogue thruster in the process. It worked, but now he was left with just the small reentry jets in the rear to steer with. He brought the ship back under control, and found that he had used up most of the fuel. He and Scott had to make an emergency splashdown, but senior mission managers noted the calm with which Armstrong had addressed the crisis.

Astronaut Grissom knew the Gemini better than anyone, and he always felt that the ship was a calculated risk. In a somber moment shortly before his launch on Gemini III, he had told his wife, "If there's a serious accident in the space program, it'll probably be Gemini, and it'll probably be me." Only a few days earlier he had alarmed senior NASA managers by being a little too honest at a pre-flight press conference. "We're in a risky business."

Out of the limelight

While the world's attention was focused on NASA's astronauts and their Soviet competitors, an alternative space program was under way, centered on the Edwards Air Force Base in the Californian Mojave desert. An old NACA project, inherited from the late 1950s, was scoring remarkable successes. Throughout the 1960s the X-15 rocket plane, a brutish dart-shaped machine, climbed to the very edges of space then plunged back to earth and made a horizontal landing, before being refurbished, refuelled and sent up again. It was—as near as makes tantalizingly little difference—the world's first true spaceship, even though it couldn't quite achieve orbital velocity. If only time had allowed, the X-15 could have been the ancestor of all subsequent American manned spacecraft. But the impetus of the moon pushed NASA towards simpler capsule designs.

Other and even stranger vehicles were tested in the desert, as some NASA engineers shied away from the rather embarrassing spectacle of capsules splashing down into the ocean and astronauts being hauled to safety by helicopters as though they were being "rescued" rather than "recovered".

The M2-F2 lifting body was one of several near-wingless landing vehicles pioneered by NASA in the 1960s. In this photo, an F-104 pursuit plane monitors the lifting body's progress as it comes in to land. Bruce Peterson's crash in May 1967, below, was a reminder that such radical experiments were hazardous.

NASA engineer Dale Reed championed a different and more aerodynamic kind of capsule known as a "lifting body".

As he recalls, "Senior managers didn't believe a capsule without wings could fly to a touchdown on land, so they splashed 'em down in the sea. We built a wooden prototype of our design for $10,000, essentially small change that we didn't have to get approval from head office to spend. Then we towed the vehicle behind a car, a souped-up Pontiac. The first pilot, Milt Thompson, controlled it beautifully."

By the summer of 1963 the wooden lifting body M2-F1, nicknamed the "Flying Bathtub", was reaching ever higher into the air, towed behind an aircraft and then released. NASA and the Air Force were pursuaded to fund a fleet of sophisticated metal lifting bodies capable of high-speed high-altitude flight. They were carried 60,000 feet into the sky, tucked beneath the wing of a B-52 bomber. And then they were dropped. Their pilots had only one chance to touch down safely... As with all high-speed aerospace experiments, there were unknown hazards ahead, both for the X-15 (which claimed one fatality) and the lifting bodies.

Neil Armstrong opposite, proved his credentials as a test pilot aboard the X-15 rocket plane. It was an extremely fast and powerful machine. Inevitably it could be temperamental. The wreckage caused no fatalities, but on 15 November 1967, test pilot Major Michael Adams was killed when his X-15 spun out of control and disintegrated high in the atmosphere, at five times the speed of sound.

Bruce Petersen, a squat but solid man in his late sixties, looks very dashing with his black eyepatch, brown flying jacket, chinos and boots. As he scans the flat dry lake bed that makes up the flight line at Edwards Air Force base, he describes his crash in a fully equipped metal version of Dale Reed's lifting body, dropped from 40,000 feet, with a real sense of spice and drama. With his laconic recollections of bravery and catastrophe at high speeds, he is the very embodiment of "The Right Stuff". He beams a wicked smile and tells his story: "It was May 10 27, 1967. What happened was the stand-by rescue helicopter pilot didn't attend the briefing on the morning of my flight and he got in the way of my ship. I lost control because this was the first flight of an untested design, not quite the same shape as the wooden prototype that had done so well previously, and this thing didn't fly too well. I got into a severe roll motion, and on top of that I was busy on the radio, saying, 'Get that chopper out of my way.' Maybe I lost concentration for a moment, and I was late putting the landing gear down by about one and a half seconds. I hit the floor and rolled over several times and the ship ended upside-down. And that was pretty much the end of

the ball game." Archive footage shows this, and later, the rescuers standing around with their hands in their pockets gloomily surveying the wreckage in the certain knowledge that Petersen must be dead.

According to Petersen, they eventually realized he was still alive and cut his battered body out of the wreckage. That's how Peterson became the inspiration for the popular 1970s TV adventure show *The Six Million Dollar Man*. Footage of his spectacular crash featured in the opening credits of every episode. Lee Majors, the fictional character in that show, was half-man half-robot. Real life is more subtle. The latest robotically navigated spacecraft are coming close to making Peterson and his kind redundant. "They say the time is coming when they won't need pilots at the front end of a ship. The history of aerospace tells you that a lot of accidents have been avoided by having people in charge, and a lot of robot-controlled machines have crashed. I crashed too, but I was able to explain why, and that meant we could fix the lifting bodies and get them right so that the next guy wouldn't crash. But the robotic trend is unstoppable, and that's the way we are going."

Lifting body pilots, above, did not enjoy as much public adulation as astronauts. Veteran X-15 pilot Milt Thompson in front of a replica X-15 in 1993, opposite.

USAF
APU EXHAUST
APU EXHAUST

03 THE ALTAR OF APOLLO

"In 1969 I predicted that the Saturn V moon rocket would dominate the 1970s like a colossus. In the event, far from dominating the '70s, the Saturn was dominated by them. But the time will come when Project Apollo is the only thing by which we will remember the United States, or indeed the world of our ancestors, the distant planet earth."
Arthur C. Clarke, 1995

THE ALTAR OF APOLLO

> Historians still debate the role played by John F. Kennedy in shaping NASA's fortunes, but rather less attention is devoted to the man who actually led the agency in those years. It was not Wernher von Braun, despite his great celebrity and skilled leadership of the Saturn V's development. The man in charge of NASA was a somewhat unlikely figure, a burly, talkative bureaucrat with little experience of space or rockets. What he understood was politics.

When President Kennedy took office in 1961, NASA was still a relatively minor arm of the federal government. The White House advisors made contact with at least seventeen distinguished scientists and business leaders, asking if they might like to head the agency, but none of them viewed the job as a satisfying career move. In desperation, Vice President Lyndon Johnson strong-armed an unlikely candidate, a lawyer from North Carolina, into accepting the post.

James Edwin Webb was born in October 1906, in the small town of Tally Ho, North Carolina. As a young man, to help his family get along he took classic small-town jobs: working on farms, helping out in nickel-and-dime stores, truck-driving on construction sites. His father was the superintendent of schools in Granville County, and his mother was equally concerned about education and social progress. The highly intelligent young man grew up as a progressive Democrat idealist. After the Wall Street Crash of 1929, career opportunities for a young man in a small North Carolina community were limited, so Webb joined the Marine Corps. A handful of young men were selected to be part of the first aviation squad in the Corps, based on Long Island, New York. Webb's squad

UNITED STATES
USA
USA

Page 95, a Saturn V rocket is prepared for the first unmanned test of the system in November 1967.
Opposite, an aerial view of NASA's huge Johnson Space Center alongside Clear Lake, near Houston, Texas.

consisted mainly of privileged Ivy-Leaguers, yet he was readily absorbed into the camaraderie of the group. He found he had the social skills to blend into any stratum of society, and he discovered that he could compete and win against richer people. It confirmed his Democratic view of America as a nation where access to opportunity could help people overcome the barriers of class and wealth. Meantime, he knew it could do no harm to better his material career prospects. He chose a classic route—studying law.

On a trip to Washington D.C. to catch up with education officials he had known back in North Carolina, Webb was introduced to a congressman who just happened to be in urgent need of a new secretary.

This was the era of Franklin D. Roosevelt's New Deal, a bold attempt to get America moving again after the Depression years using vast sums of tax money to initiate building projects and infrastructure development. Webb thrived in this idealistic environment. He also turned out to be a smart political operator, and a successful businessman.

In 1938, now happily married, he won a place at a Washington law firm. His experience in both law and aviation then caught the attention of the executives of the Sperry Corporation in New York city as they geared up for the coming war. They recruited him. His imput helped shape Sperry into one of the most successful defense contractors of the time, before leaving them in 1944 for a tour of duty with the Marine Corps as a training officer.

After the war, Webb returned to public service, rising to the post of Director of the Bureau of the Budget, a position he held until 1949. President Truman then asked him to serve as Under Secretary of State. But when the Truman administration ended in 1953, Webb didn't like the idea of working for the incoming Republican president, Dwight D. Eisenhower. He left Washington for a senior position at Kerr-McGee, the giant oil and nuclear power company in Oklahoma City.

And there, his contribution to the greater sweep of history might have come to an end. He could have finished his career as a talented, rich and successful businessman just like many others, albeit with some flair for charitable works and public service: the embers from his youthful New Deal fires not yet burned out. Today, we might not have had much cause to remember him in any broader context. But history wasn't finished with Jim Webb.

On January 17, 1961, as Eisenhower prepared to cede power to a new Democrat president with ideas very different from his own, he made an extraordinary farewell address to the nation in which he reminded American citizens that a gigantic armaments industry had been created since World War II where none had existed previously, and it was threatening to change all of society: "In the councils of government, we must guard against the acquisition of unwarranted influence, whether sought or unsought, by the military-industrial complex. The potential for the disastrous rise of misplaced power exists and will persist. We must never let the weight of this combination endanger our liberties or democratic processes." He also suggested that "public policy could become the captive of a scientific-technological elite".

Eisenhower surely would have been dismayed if he had known that the newly formed NASA would soon be under the control of a man who believed that the "scientific-technological elite" was not a threat to America's future, but the solution to all its ills.

Webb believed that American industry was not meeting the challenge posed by Soviet strength. He also felt that the colleges and universities were not creating a sufficient number of graduates qualified for the technological future that lay ahead.

"You see, you've got the physicists here and the astronomers here, and the chemists over there—somebody in the government has got to tie it all together," he said. And that somebody might as well be Webb.

In 1961, Johnson and Kennedy asked Webb to lead NASA because they could not find anyone else. As a fellow southern politician, Johnson got along fine with Webb, but there is no reason to suppose that Kennedy and his team were especially fond of him. If they had known how important space was about to become, they would undoubtedly have appointed someone more to their liking. As the political writer Charles Murray observed in a fine 1989 history of Apollo:

"Stocky and voluble, Webb at fifty-five was of a different generation than most of the others in this new administration, and from a different world. Instead of Harvard and wire-rimmed glasses, clipped accents and dry wit, he was University of North Carolina and rumpled collars, corn-pone accent and down-home homilies, a good ol' boy with a law degree."

While many NASA insiders thought that the great new space adventure announced by Kennedy must surely continue for generations to come, Webb understood that such a fragile, fleeting moment in history was a rare window of opportunity that would surely close as suddenly as it had opened. Moving with a speed and decisiveness that seem unimaginable today for any government bureaucrat, he consolidated NASA's influence across a vast crescent-shaped swathe of coastal states: California, Texas, Mississippi, Louisiana, Alabama, Florida, Virginia, Maryland and Ohio, the one inland state. Webb spent most of his time at NASA headquarters in Washington D.C., from where he could keep a close eye on the politicians.

A sensible way to run a rocket facility might have been to put the launchpads, control centers and construction hangars all in one location, and certainly no one doubted Webb's reasons for expanding a small Air Force missile testing station at Cape Canaveral, on the Florida coast, into a giant moonport. But why build the mission control center so far away, in Houston, Texas, of all places? And why build the Apollo capsules in California, booster components in Alabama and Louisiana, and yet more huge chunks of equipment on Long Island? His scatter-gun dispersal of NASA operations and private industrial contracts angered many people, not least the Congressional representatives of those states where NASA contracts were not so thick on the ground. They saw the distributions as politically motivated, perhaps even unethical. But Webb was almost impossible to intimidate. "I made up my mind early in the game that I couldn't let anybody dictate decisions that were at a technical level, whether it was the President, the Vice President or the scientists."

James Edwin Webb, NASA administrator for the Apollo moon project from February 1961 until October 1968.

"We choose to go to the moon in this decade, and do the other things, not because they are easy, but because they are hard, because that goal will serve to organize and measure the best of our energies and skills, because that challenge is one that we are willing to accept, one we are unwilling to postpone, and one which we intend to win."
President John F. Kennedy, September 12, 1962

He was playing a clever game. First, he wanted to ensure that as few Congressional representatives as possible would wish to block NASA's funding, for fear of losing space jobs in their home states. Second, as an old-style Democrat he believed passionately that one of the proper purposes of any large government contract should be to stimulate employment and education opportunities as widely as possible across the nation as a whole. While NASA's own workforce grew to 30,000 people in the mid-1960s, private contractor staff amounted to more than ten times that number. That could not last, but for a few years at least, Webb's large-scale ambitions delivered jobs and a sense of self-worth and patriotism to nearly half a million space workers, from rocket scientists, computer experts and metal benders to building contractors, secretaries and caterers. He created real-estate booms in Florida, California, Texas, Alabama, Virginia and other states besides, as the various NASA "field centers" grew from modest laboratories into complexes the size of small cities, with sprawling suburbs. These were vanguard projects (or so Webb and his allies hoped) in a new space-age economy.

Webb also fought behind the scenes to ensure that the Apollo lunar project alone would not define NASA to the exclusion of everything else. The less glamorous scientific programs were equally important, he believed. We take this multifaceted character of space exploration for granted today, but in the feverish atmosphere of the 1960s moon race it had to be defended. President Kennedy made many eloquent speeches about the benefits of Apollo, and steadfastly supported NASA in public even while privately telling Webb that the estimated costs of reaching the moon appalled him. "I'm not that interested in space," he admitted bluntly at a White House meeting in November 1962. "We're ready to spend reasonable amounts of money, but we're talking about these fantastic expenditures, which wreck our budget on all these other domestic programs, and the only justification for it, in my opinion, is because we hope to beat the Russians!"

Kennedy wanted a solid assurance: did Webb think Apollo was NASA's number-one priority? "No sir, I do not," he replied. "I think it is one of the priorities." He argued that America needed a broader range of activities in space. Onlookers were amazed as a shouting match threatened to brew between the two men, but Kennedy was apparently swayed and Webb's view prevailed. Throughout his tenure at NASA he fought to protect and promote unmanned scientific and astronomy programs that we now celebrate as among the agency's most valuable achievements: probes to Mars and Venus, and the first tentative steps towards exploring Jupiter, Saturn and worlds beyond.

NASA Associate Administrator Robert Seamans, Marshall Space Flight Center director Wernher von Braun and President John F. Kennedy inspect some of NASA's rapidly expanding facilities in September 1962.

S-IV
UNITE

So where did Webb acquire his particular set of ambitions? A generation of politicians steeped in the New Deal era of the 1930s, world war in the 1940s and nuclear competition in the 1950s, had grown accustomed to the idea that nations as a whole could be run like cohesive machines. The perfectible society, led by selfless philosopher-kings, is one of the oldest ideals of politics, and it entranced many governments across much of the twentieth century. Already, in these modern times of untrammeled free-market economics, we have forgotten the extent to which America once operated a vast "command economy." Just as in Soviet Russia, the overriding belief was that government influence over the new forces of science and machinery was crucial to a nation's success.

In the 1960s, American industry was coasting along selling outdated consumer goods. Skilled marketing exploited mass psychology to persuade Americans to buy products as never before. The economy was incandescent with success, yet the products themselves weren't necessarily based on genuinely new technologies. Private industry simply wasn't looking at the "Big Picture," Webb argued. The quest for short-term corporate profits couldn't satisfy the longer-term economic requirements of society as a whole in a technological era. Perhaps even more seriously, private industry alone couldn't ensure technological supremacy on the military front. This was an overwhelming problem now that Soviet Russia was apparently at its zenith. "Thinking Big" was obviously the government's job. Massive federal expenditure on high-tech scientific research and development was initiated with the aim of filtering innovations into the private sector.

Roosevelt had championed technocratic Big Thinking to salvage an ailing nation in the 1930s; wartime mobilization made it mandatory; Eisenhower had grave doubts in an era of 1950s post-war complacency but continued to sanction it; Kennedy took it pretty much for granted as part of his "new frontier"; and Lyndon Johnson tried to keep its dying embers alive when he became President, even as the war in Vietnam crippled his budgets and broke his morale. Presidents Nixon and Carter slammed on the brakes, but in the last decade before the demise of Big Thinking, NASA's Jim Webb became one of its most committed champions.

Opposite, something of the vast scale of a Saturn V is conveyed by this shot of an engineering test version inside the cavernous Vertical Assembly Buildng (VAB) at the Kennedy Space Center in 1967.

At first, space exploration had seemed nothing more than a game for Buck Rogers cadets, and not a sensible aim for national policy. Webb, Johnson and Robert Kerr (their powerful ally on the Senate Space Committee) believed that the rocket's industrial impact on the ground was just as significant as anything it could achieve in space.

No American technological enterprise since the construction of the Hoover Dam in the 1930s better exemplified Big Thinking than NASA in the 1960s. Secret military programs may at times have benefited from even more lavish spending than NASA, but it was the sheer concentration of NASA's funds towards one particular target—the moon—that lent it such visible grandeur. Even so, Webb saw the lunar landing as a means rather than an end. He came to embody the twentieth century technocrat in full glory: spending vast sums of taxpayers' money on science and machinery in order to improve education, stimulate flagging industries, and redefine the management structures and beliefs of large-scale organizations across the country. Webb's vision for Apollo was nothing short of a model for grand social engineering. It wasn't the moon so much as a transformation of the entire country that he and his kind were after. He campaigned throughout these years for a new brand of "space-age management" in many areas of national life, not just in rocket development.

According to the Smithsonian Institution's chief space historian Roger Launius, "Webb thought it was possible to create the perfect organization. He talked

UNITED
STATES

about the application of NASA techniques to other sorts of activities such as homelessness and poverty."

As a manager, Webb believed that people functioned best when they felt responsible for their own work. He simplified the layers of middle-management in NASA and created a "flat" organization where junior staffers could talk directly to senior managers. Many commentators have identified one of the distinguishing features of both the Challenger and the Columbia shuttle accidents as the profound difficulty experienced by ground-level engineers when trying to alert senior managers to safety problems. The degradation of Webb's management structures has been mourned by many NASA watchers. He trusted and supported his colleagues in the agency and sheltered them from external interference. Between 1961 and 1967, NASA exhibited perhaps the most successful cutting-edge management of complex engineering and manufacturing that the world has ever seen. The "can-do" spirit was born in those years under Webb's particular style of management. But was it real or just a dangerous illusion? Once again we have to keep in mind that Webb was only the tip of a vast pyramid of human endeavor. The day-to-day running of most NASA affairs fell on the shoulders of his deputy, Robert Seamans. Webb spent most of his time handling the politics, not the engineering. Until illness called him away, Hugh Dryden was NASA's wise elder counsel and co-administrator. Other legendary engineer-leaders became familiar to the public. Robert Gilruth, George Mueller, Sam Phillips, Joe Shea and Rocco Petrone, with thousands of other dedicated people, made NASA's early successes possible. And von Braun of course played his essential part, heading up the development of the Saturn V launch vehicle.

Above, the VAB under construction, accompanied by three mobile launch towers, and opposite, Apollo 12's Saturn V emerges from the VAB in October 1969.

UNITED STATES

"The biggest and most ambitious program of all—Man's flight to the moon—was in deep and perilous trouble, and Congress was unaware of that fact."
Senator Walter F. Mondale, 1967

Apollo 1

> By the time the Gemini project had wound down in 1966, NASA's reputation was so high it was heralded around the world as one of the greatest organizations that ever existed. Webb and many of his colleagues were tempted to believe in their own myth. Such a peak of vanity could only be a place from which to fall. NASA was about to discover that it was no more immune from human error than any other arm of the American government. Apollo was placed in jeopardy by scandal even before its first mission had flown.

In the early evening of January 27, 1967, mission commander Virgil Grissom and his crewmates Edward White and Roger Chaffee clambered into the first flight-ready Apollo command module, atop a half-sized version of the Saturn rocket, the 1-B. This was to be a routine checkout procedure. They would run a simulated countdown with all systems running, but would not actually ignite the Saturn's engines for take-off. There was no fuel in the rocket's tanks. The goal, for now, was simply to check out the Apollo's systems.

As Chaffee climbed through the hatch, he complained that the capsule's interior smelled of sour milk. Then there was a glitch in the radio links. Two ground controllers, one nearby at the Kennedy launch center and another at mission control in Houston, were having trouble hooking up.

"Do you copy?" asked one controller.

"No, I didn't read you at all. I can't read you. You want to try the phone?"

Inside the capsule, Grissom was furious. "How are we going to get to the moon if we can't talk between two or three buildings? I can't hear a thing you're saying."

"Can you guys talk together up there in the command module?"

"I said, how are we going to get to the moon if we can't talk between two or three buildings?"

The mood was tetchy as pad technicians locked Apollo's heavy hatch into place, sealing the crew inside. Five hours into the test, just after 11:30 at night, Grissom's garbled voice on the crackling radio link said, "We've got a fire in the capsule." A few seconds later another voice, possibly White's, was more urgent. "Hey, we're burning up in here!" There was a brief yell, and then just a hiss of static as the radio went dead.

Suddenly the side of the capsule split open, and the top of the launch gantry was engulfed in acrid smoke and flames. The pad crew tried desperately to get the astronauts out. The smoke was impenetrable and the heat overpowering. It took four minutes to open the Apollo's hatch. By then the astronauts were dead.

Apollo 1 astronauts Virgil "Gus" Grissom, Ed White and Roger Chaffee pose for a publicity photo in front of the launch pad.

NASA
NASA

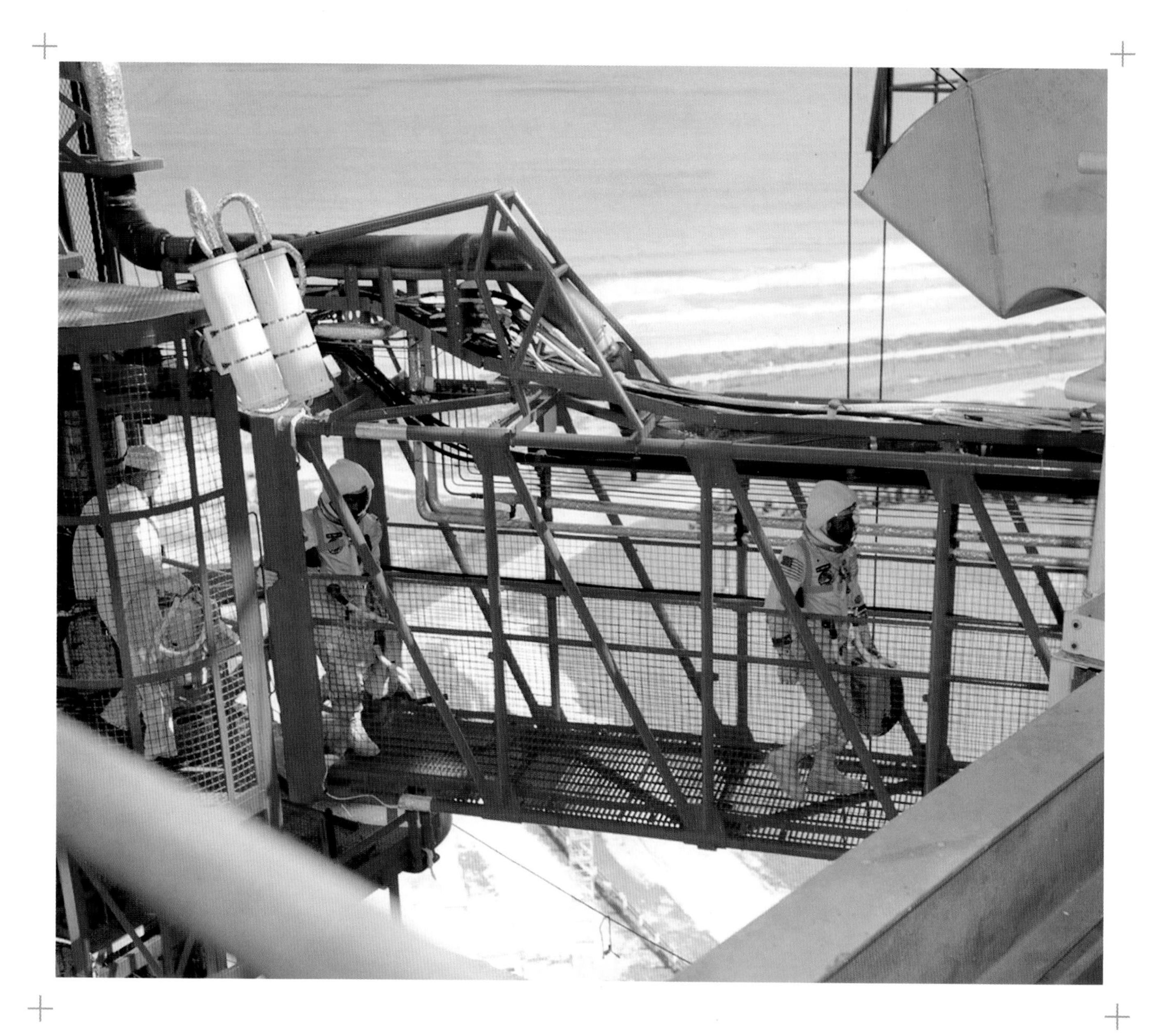

Above, the Apollo 1 astronauts walk across an access gantry near the top of the lauch tower, heading for their flawed capsule. Opposite, NASA deputy administrator Robert Seamans, his chief James Webb and senior Apollo manager George Mueller face questions from an angry Congress.

NASA investigators found to their dismay that many fire risks had been ignored or overlooked, such was the mood of confidence at that time after the brilliant successes of the Gemini program. Some of the blame for the fire was laid at the door of North American Aviation, the California-based company most responsible for building the capsule. Faulty wiring caused a tiny short-circuit spark. Ordinarily that should not have been a disaster, but NASA had chosen to pump pure oxygen into the Apollo's cabin for the astronauts to breathe. It was pressurized at higher than normal levels so that the capsule would not be contaminated by dust and moisture from the Floridian air. Oxygen is highly flammable under pressure, and somehow NASA missed this obvious danger.

Webb felt that his dream of a perfect technological enterprise had unraveled. No matter how excellent his system of management had been, it could not protect against this most unexpected of disasters: a capsule on an unfuelled rocket catching fire and exploding while still on the ground. His Washington enemies took advantage of a high-profile Congressional enquiry to attack him as a secretive manipulator who treated NASA as his personal fiefdom. He felt morally crushed, but tried not to show it. He told the companies working on Apollo that he would cancel their contracts if they didn't revise their working methods and produce better hardware. It was a dangerous gamble: if he had actually followed through on his threat, Apollo almost certainly would not have reached the moon "by the end of the decade", President Kennedy's famous deadline. It is now clear that Russia might have achieved some kind of manned lunar mission given a little more time: not a landing, but a lunar circumnavigation in a one-man variant of their Soyuz ship. This would have been quite enough to take the wind out of Apollo's sails.

After the Apollo 1 fire, newspaper and TV reporters zeroed in on the complex business dealings between NASA and America's powerful aerospace manufacturing companies. Webb suffered a humiliating trial by innuendo and rumor, and he resigned from NASA in 1968 just before the first successful manned launch of an Apollo 7 craft for an earth orbital test. The reasons he gave for stepping down were perfectly honorable: if Richard Nixon won the White House, as seemed very likely, then Webb wanted a new space chief, Tom Paine, to take credit for Apollo in this new Republican era, without being tainted by past Democrat scandals.

"The space program is probably the most centralized government spending program in the United States. It concentrates into the hands of a single agency full authority over an important sector of our economy. It could well be described as corporate socialism."
Senator William Proxmire, 1963

Lyndon Johnson greatly admired Webb, and considered him an ally. But he was not unduly disturbed to see him go, because some of the business scandals surrounding the Apollo fire had touched on other sensitive areas of Washington politics. Bobby Baker, a close confidante of both Johnson and Kerr, was enmeshed in a sensational bribery scandal—and his misdemeanors had only come to light because of his ownership of vending machines installed in some of the factories that made Apollo. One or two unscrupulous freelance lobbyists tried to exploit Apollo for personal gain. The noble arena of space was not exempted from their greed. By the time Apollo 11 made its historic touchdown on July 20, 1969, Webb's name had been all but erased from the public's consciousness. Yet he was the man who prepared NASA for its assault on the moon and warded off huge political dangers in the wake of the Apollo fire. In so doing, he laid the foundations for the agency as we know it today: politically controversial, dogged by seemingly endless troubles, yet still the world's leader in the exploration of space.

As Roger Launius says, Webb had "fought, charmed, cajoled, politicked and bargained" his way through the complexities of politics and money at the highest levels, but only in the national interest. He was so honest in his personal finances he wouldn't even ride a government limousine. However, he was evasive when questioned by Congress about the fire. In particular he was caught off-guard by a critical document called the Phillips Report, prepared by other managers at NASA in 1963, in which the capsule's manufacturers were sharply criticized for their poor construction standards.

Webb's manipulative and high-handed running of NASA made him many enemies, yet he proved that a government organization could commit itself to a goal and achieve it, so long as the leadership was sufficiently sure of itself. And today, in the light of plans to return to the moon in the coming decade, we all want to know how to ensure—how to *manage*—that same kind of triumph again. On the other hand, the way in which this public servant wielded so much control over such a vast amassing of resources (at its mid-1960s peak, 5 per cent of federal spending) caused some people to regard him and his colleagues as a menace, and NASA as a dangerous institution. How ironic that as America reached for the moon in order to demonstrate the virtues of its free-market democracy over Soviet tyranny, it seemed to be embracing some of the same economic tools as its adversary. One prominent senator, William Proxmire, told Congress: "The space program is probably the most centralized government spending program in the United States. It concentrates into the hands of a single agency full authority over an important sector of our economy. It could well be described as corporate socialism."

This Devil Ship!

> A no less terrible disaster in the Soviet space program helped put the Apollo fire into its wider perspective. On April 23, 1967, the first of Russia's new Soyuz spacecraft was propped up against the gantry at Baikonur, ready for launch. Archive footage shows its unhappy test pilot, Vladimir Komarov, and some very subdued technicians. It's almost as if everybody knew that a bad day was in store. They had good reason to be anxious. Yuri Gagarin and several other cosmonauts had tried to send word to the Kremlin that the new ship was riddled with design flaws and should not be launched until it had been improved. No one in authority listened, and just as his friends and colleagues had feared, Komarov hit trouble as soon as he reached orbit. One of the solar power vanes on the rear equipment module of the Soyuz refused to deploy, and his guidance computers ran short of power. Launch of a second Soyuz that was supposed to rendezvous with Komarov was cancelled while ground controllers worked to fix his power problems. After 18 orbits (26 hours), mission controllers decided to bring the capsule down to earth as soon as possible. Komarov couldn't properly line up his capsule for reentry, and he complained: "This devil ship! Nothing I lay my hands on works properly!"

After Komarov's hectic reentry, a small drogue parachute deployed, but failed to pull the bigger 'chute from its storage bay. A back-up parachute was released and became entangled with the first drogue, probably because the capsule was still spinning out of control. There was nothing to slow the descent, and Komarov slammed into the ground with all the force of a 2.8-ton meteorite. He was killed instantly. NASA and its astronauts sent messages of sympathy, just as the Soviets had done for the Apollo 1 crew. Congressional snipers could no longer suggest that NASA had been uniquely or unusually careless in its preparations. Space exploration was a risk for both sides, and especially so in the febrile atmosphere of a race.

The main trouble for Russia's space program was that it had lost its guiding hand. On January 14, 1966, Korolev checked into the Kremlin Hospital for a supposedly routine intestinal operation. Weakened by years of ill health and overwork, his body was far more fragile than the doctors had suspected. Internal bleeding proved difficult to control, and two huge cancerous tumors had developed in his abdomen.

After a lengthy operation Korolev's heart gave out and he died at the age of 59.

"Something to remind me of life"

Before his death, Korolev had been planning a giant lunar booster to put a Soviet man on the moon. Now that he was gone, the space effort became a shambles, with two dozen bureaus fighting for dominance. For a while, the Chief Designer's legacy was put in the hands of a much weaker figure, Vasily Mishin, but Korolev's old rival Glushko scavenged for power and eventually took control of much of his old empire, with obvious satisfaction. Glushko had a big ego and undoubted engineering talent, but he wasn't too good at managing people. In fact, no one seemed to have the political skills or strength of character to run the space program properly, and this was a great shame for the Soviet Union. If their Chief Designer had lived a few years beyond 1966 he might have given NASA's Apollo some real competition.

Despite appearances, Korolev had never really recovered from his cruel Siberian imprisonment three decades earlier. Throughout all his years working to give the Soviets a lead in space, he seldom discussed his arrest and beating under the old Stalinist regime. People thought of him as a burly man built like a bear, yet the truth was that his body had been made rigid by countless ancient injuries. He couldn't turn his neck but had to swivel his upper torso to look people in the eye, nor could he open his jaws wide enough to laugh out loud.

Two days before he was scheduled for surgery he was resting at his home in the Ostankino district of Moscow. Gagarin and fellow cosmonaut Alexei Leonov came to visit him with several other friends, and at the end of the evening, just as most of the visitors were putting on their greatcoats to leave, Korolev said to his two favorite cosmonauts, "Don't go just yet. I want to talk." His wife Nina fetched some more food and drink, and for four hours, well into the early hours of the morning, Korolev told the story of his arrest: a story that Leonov has never forgotten.

"He told us how he was taken away and beaten. When he asked for a glass of water they smashed him in the face with the water jug. They demanded a list of so-called traitors in the rocket laboratories, and he could only reply that he had no such list." Korolev described how he saw, through puffy eyes, that his captors had pushed a piece of paper between his bruised fingers for him to sign, how they beat him again and sentenced him to ten years' hard labor in Siberia. He was only recalled to Moscow when an old ally of his, the renowned aircraft designer Andrei Tupolev (recently a prisoner himself), specifically requested him for war work. He was assigned to a less harsh special prison facility for engineers, which included design offices. But no special arrangements were made to transport Korolev to Moscow and so he had to improvise. Cold beyond endurance and hallucinating with hunger, he found a hot loaf of bread on the ground one day, apparently dropped from a passing truck. "It seemed like a miracle," he told Gagarin and Leonov.

He worked as a laborer and shoe repairer to earn his passage back to Moscow by boat and rail. His teeth were loose and bleeding because he hadn't eaten

"SOVIET FIRES EARTH SATELLITE INTO SPACE. IT IS CIRCLING THE GLOBE AT 18,000 M.P.H. SPHERE TRACKED IN FOUR CROSSINGS OVER U.S."

The *New York Times* main headline, October 5, 1957

fresh fruit or vegetables in over a year. Trudging along a dirt track one day, he collapsed. An old man rubbed herbs on his gums and propped up his body to face the feeble sun, but he collapsed again. He was on the verge of death. Then, as Leonov vividly remembers, "He told us he could see something fluttering. It was a butterfly, something to remind him of life."

It seems the ailing Chief Designer wanted to unburden himself after so many years' silence. This great powerhouse of a man had never spoken before in such a fragile, wounded and personal way, and the two young cosmonauts were deeply affected by what they heard. "This was the first time that Korolev had ever talked about his imprisonment in the gulag, since these stories are usually kept secret," says Leonov. "We began to realize there was something wrong with our country. On our way home, Yuri couldn't stop questioning. How could it be, that such unique people like Korolev had been subjected to repression? It was so obvious that Korolev was a national treasure."

"One day in the early 1960s, Sergei Korolev was looking at a newspaper photograph of Wernher von Braun, then being lionized in the United States for his part in the upcoming Apollo program. 'We should be friends,' he commented."

James Harford, *Korolev*, 1997

"On our way home, Yuri couldn't stop questioning.

'How could it be, that such unique people like Korolev had been subjected to repression?'

It was so obvious that Korolev was a national treasure."
Cosmonaut Alexei Leonov, 1997

Э.ЗОЛЯ
25
26
3
5
10
11
12

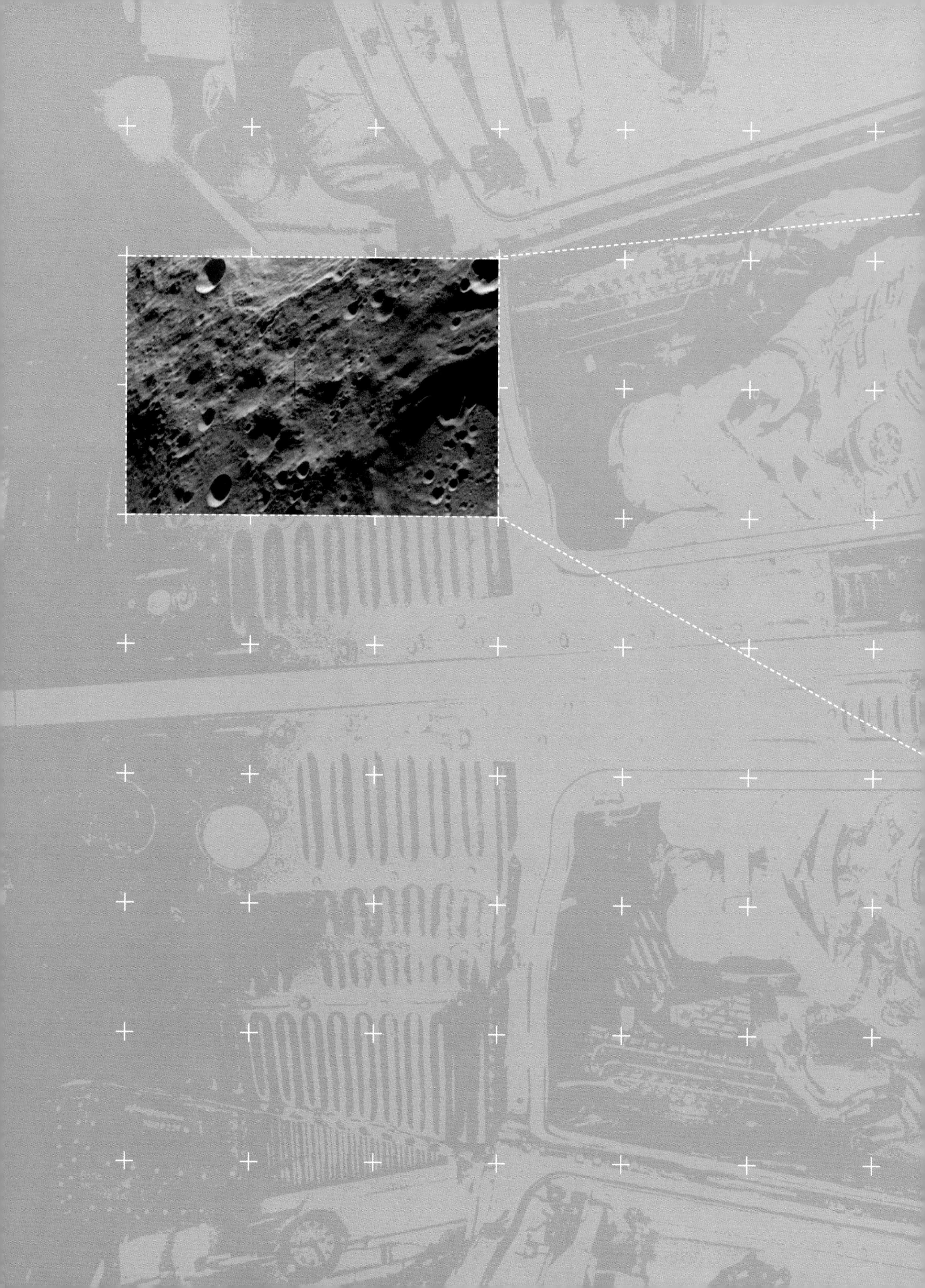

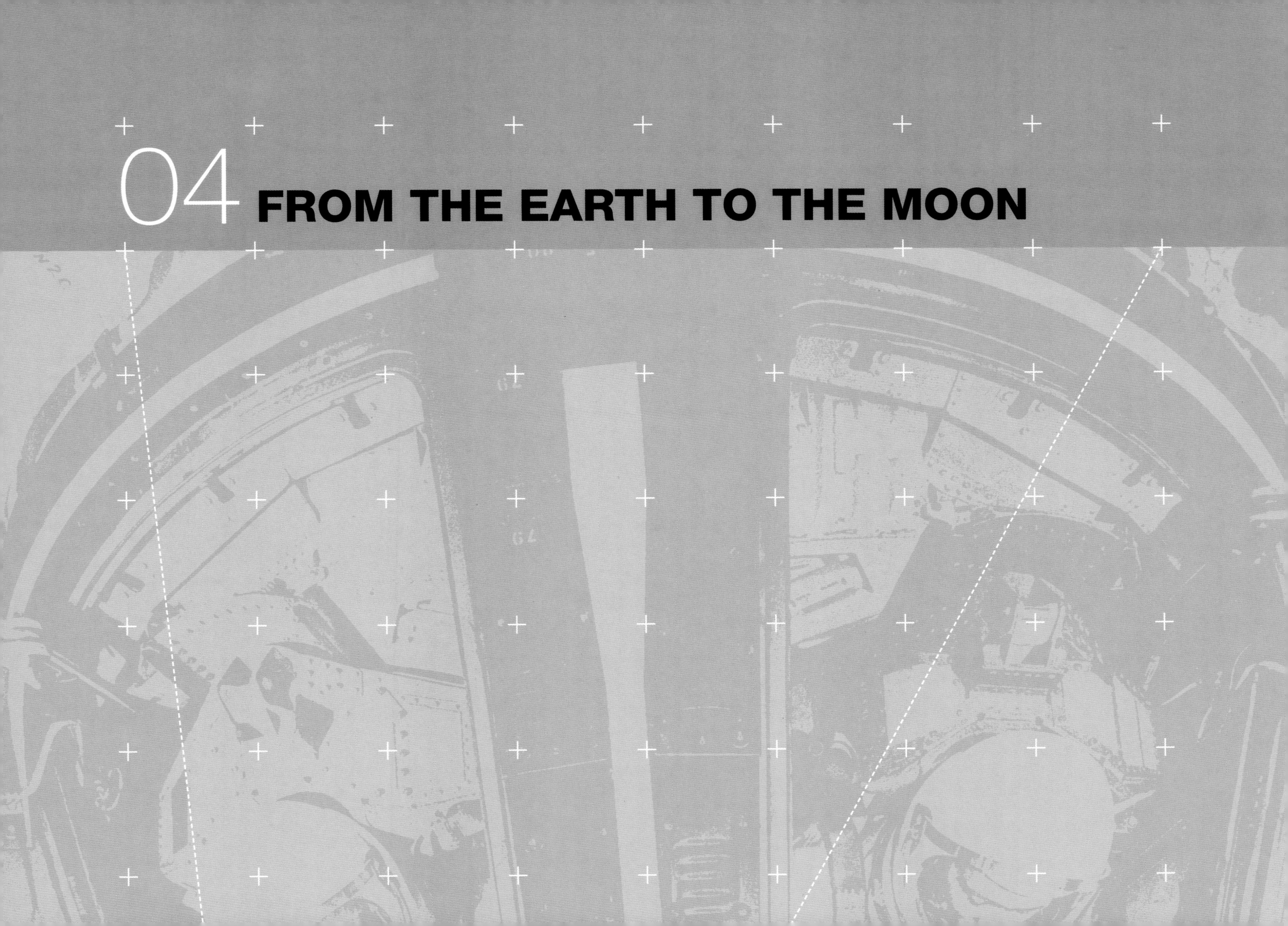

04 FROM THE EARTH TO THE MOON

"As the Saturn V moved off the pad, the sound finally reached across the marsh and slammed into the viewing area. It came first through the ground, tremors that shook the viewing stand and rattled its corrugated iron roof. Then came the noise, 120 decibels of it, in staccato bursts. People who were there would recall it not as a sound but as a physical force."
Charles Murray and Catherine Bly Cox,
Apollo: The Race to the Moon, 1989

FROM THE EARTH TO THE MOON

> Although 1967 had been an appalling year for NASA, there was one beacon of hope. By November, Wernher von Braun's gigantic Saturn V booster rocket for the Apollo was ready for its first test, albeit only with an unmanned version of the capsule on top.

When technicians at the Florida launch site had stacked all three stages of the 3,300-ton booster to its full height of 360 feet for the first time, nobody really believed in their hearts that such a monster—the size and weight of a naval destroyer standing on its stern—could really lift off and fly. As if to confirm their worst suspicions, the six-day countdown stretched to seventeen as endless failures cropped up on the launch controllers' screens. Time and again, Rocco Petrone, Chief of Launch Operations, tried to get the countdown running smoothly. The fuel pumps failed. Batteries on the second stage's control system failed. The computers crashed. It seemed impossible to get all the thousands of systems working in harmony. Perhaps this monster was just too complicated to launch? An exhausted Petrone muttered: "Can we ever get this baby to go? Can we ever get all the green lights to come on at the same time?"

At last, in the early morning of November 9, the world's largest and most powerful rocket lifted off successfully on a mission designated Apollo 4. NASA announced that Apollo was back on track, at least as far as the launch vehicle was concerned. But the Saturn V still had some frightening surprises in store.

On April 4, 1968, a second unmanned Saturn V took off. From the press stands 5 miles from the pad, everything looked wonderful. As it turned out, few

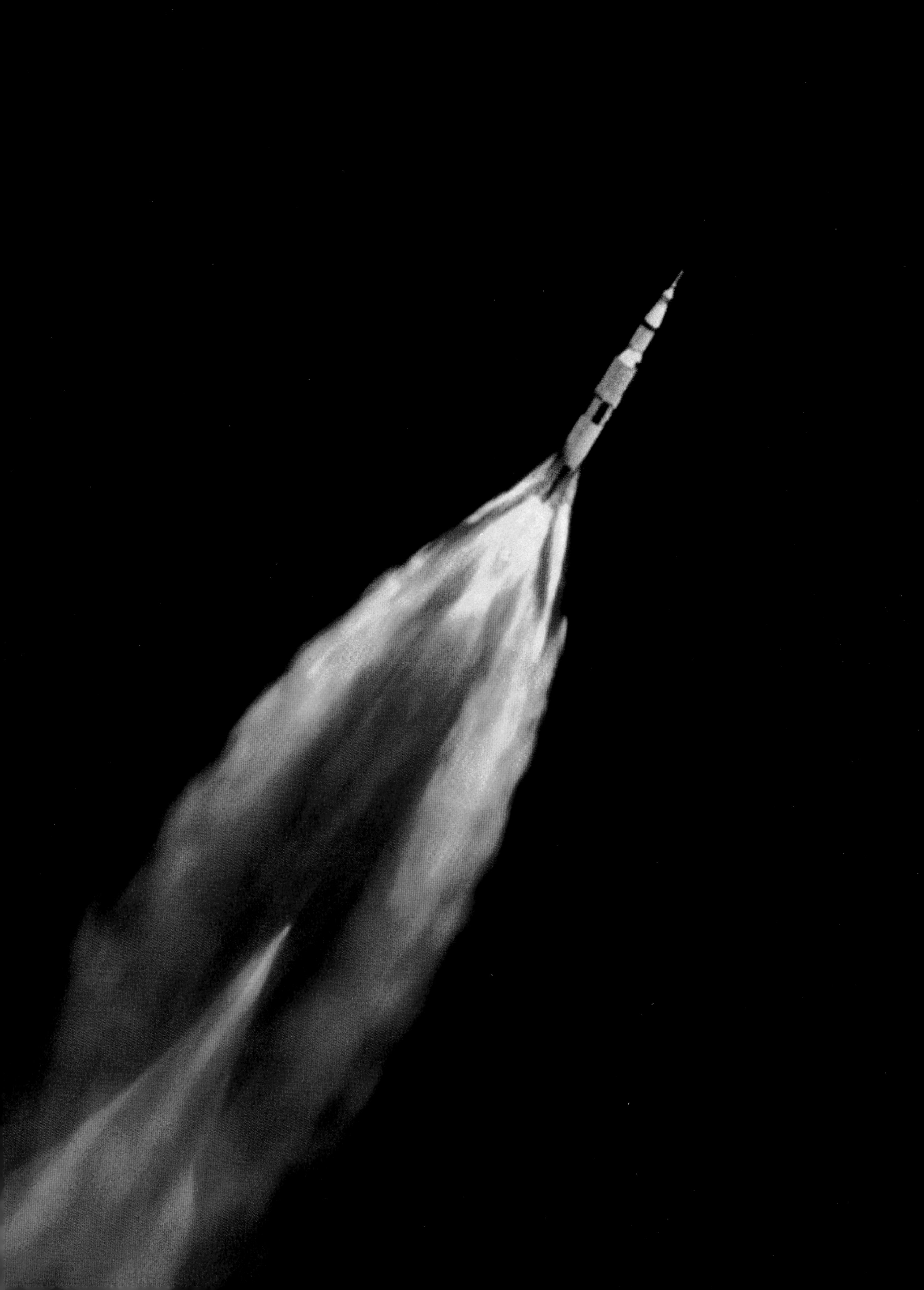

"Magnificent desolation."
Buzz Aldrin's description of the Moon, July 1969

This view of the moon, seen from the window of an Apollo lunar module, also shows the hatch of the docked command module.

newspapers or TV stations reported the launch in any detail because an earthly tragedy turned their attention elsewhere. Civil rights leader Martin Luther King was shot dead that same day.

If journalists had not been so understandably distracted, they might have asked a few questions about this second Saturn V flight. Two of the five F-1 main engines failed, leaving the rocket dangerously unstable; and at one point, too high in its trajectory for civilian observers to notice, the rocket keeled over and headed nose-down towards the Earth before leveling out again. The upper stage, with its unmanned Apollo capsule, just about made it into orbit, but ended up flying backwards around the Earth. The capsule was exposed to such heavy vibrations that many of its control systems failed. The flight was a catastrophe and if there had been astronauts aboard they might well have perished.

In the light of all this bad news, a senior Apollo manager, George Low, made an incredible decision. Next time the Saturn lifted off it would carry a crew. And it would take them all the way to the Moon.

When Jim Webb first heard about Low's plan, he exploded: "Are you out of your mind? You're putting our agency and the whole Apollo project at risk!" However, as he handed control to Tom Paine, he acknowledged that such a risk had to be taken. America could not be allowed to lose this race. NASA's new chief quickly authorized one of the greatest gambles ever taken in space.

American intelligence agencies knew that Russia's massive N-1 lunar landing booster was no great threat. Conceived by Korolev, but inefficiently pursued after his death, it was not a sound piece of engineering. According to a National Intelligence Estimate document of March 2, 1967, "Several factors militate against the Soviets being able to compete with the Apollo timetable. Their lunar launch vehicle will probably not be ready for test until mid-1968, and even then we would expect to see a series of unmanned tests lasting about a year to qualify the system before a lunar landing might be attempted."

But a smaller Soviet rocket presented more of a challenge. The Proton, developed by Korolev's rival Vladimir Chelomei, and still in use today, was ready to fly. It was possible—just possible—that Russia might launch an improved version of the Soyuz around the moon by the end of 1968, even if the capsule had to be so crammed full of extra fuel and oxygen that perhaps only one cosmonaut could fit inside. If this mission succeeded, it would be yet another Communist triumph: first satellite in space, first animal, first man, first spacewalker, first double launch, first woman, and now, the first to the Moon.

NASA had little choice but to commit its flawed Saturn V to a lunar mission as soon as possible. The spider-like lunar module landing craft wasn't yet ready to make a flight, and the only short-term option was to send an Apollo capsule to the moon without the lander. First, however, von Braun had to find out why his Saturn V had failed.

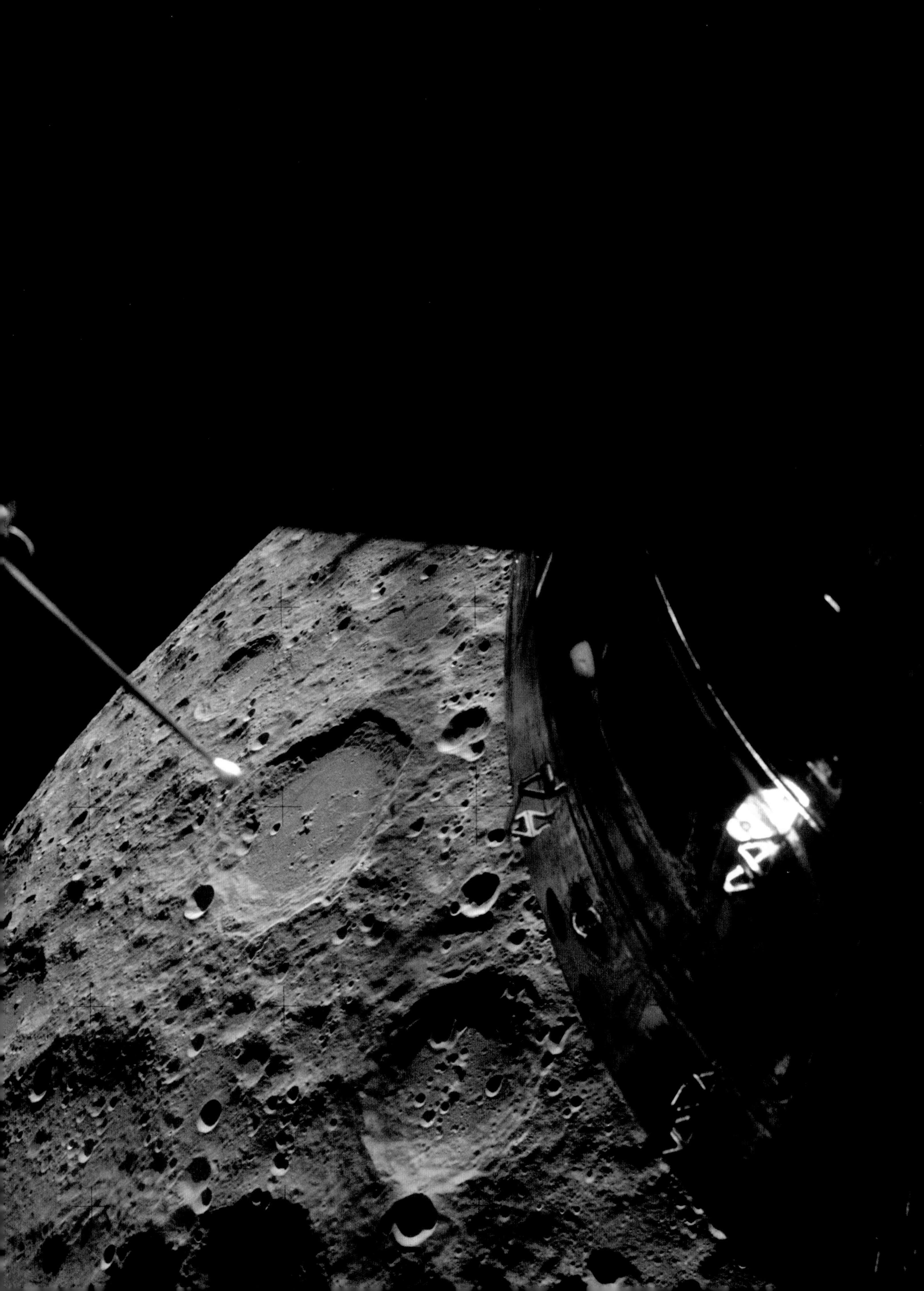

"Throughout the years I spent training for missions, my focus had been on the Moon, as well as the astral path that would take me there. Yet the first time I flew in space it was the sight of our own planet from afar that held me in both joy and awe. Looking back at our blue globe from such a distance profoundly changed my vision of space and time."
Cosmonaut Alexei Leonov, 2004

A thousand engineers from NASA and its contractors were unleashed in an emergency "crash program" to solve the problem. It turned out that kerosene and liquid oxygen thrumming at high speed through the engines had set up a resonance that shattered the fuel lines. An adjustment to the metal alloys in the pipework cancelled all the vibrations. Ground tests confirmed that the fix had worked, but it was still an incredible risk to launch the next Saturn V with men aboard.

On September 14, Russia flew an unmanned capsule, codenamed Zond, around the moon and returned it to Earth. There were many problems yet to overcome before a cosmonaut could achieve the same feat, but from the perspective of NASA, watching at a distance and with only partial knowledge of the details, this flight was extremely worrying.

Launch operations chief Rocco Petrone came under pressure to launch a Saturn towards the moon by the turn of the year. He told senior Apollo manager George Low: "That next Saturn's not going to launch before February 1969. I don't care if you give it to God!"

NASA didn't give it to God. They gave it to the next best thing: Frank Borman, one of the most respected astronauts and a candidate to make the first moon landing attempt sometime in 1969. He and his crewmates Jim Lovell and Bill Anders had been training intensively on lunar module simulators and now NASA chiefs were telling them they had to go to the moon straight away, and without a lander. It seemed a let-down, but Borman understood the importance of this unexpected mission, and agreed to fly it despite the risks. Anders, busy training as a lunar module pilot, was even more disappointed but also agreed. In NASA, the unspoken rule has always been that astronauts take whatever missions they can get, and count themselves lucky to fly at all.

It wasn't until November 12, 1968, that NASA announced that Apollo 8 would take off a few days before Christmas and head straight for the Moon: an indication of how rushed the decision was.

On December 21, 1968, Apollo 8 launched and this time the Saturn V worked flawlessly. Then came the moment when NASA controllers informed Borman and his crew, "You are 'Go' for Trans-Lunar Injection." This moment was perhaps just as great a turning point in human history as the Apollo 11 touchdown the following year. Humans were leaving the protective realm of Earth and traveling into deep space for the very first time.

Athough the N-1 lunar booster never flew sucessfully, it was still an epic piece of technology. The Soviet authorities were so concerned with secrecy, they never released any photos. The N-1 was captured at its pre-launch prime by just a few brief snaps, including long-lens images taken by western Intelligence operatives.

Apollo 8's view of earth adrift in the blackness of space, rising over a desolate lunar horizon, is the one of the most important images from the entire space program.

Two days later and 176,000 miles from home, the astronauts sent back TV pictures of the Earth: not just the curved horizon but the entire ball of our planet, surrounded by blackness. The mono TV camera aboard the ship revealed barely more than a white blob, but it was still sensational. Jim Lovell thought out loud for his TV audience: "I keep imagining I'm a lonely traveler from some other planet, and I wonder what I'd think about the Earth from this distance. Would I think it was inhabited?"

Ground communicator Michael Collins (later to fly aboard Apollo 11) replied lightly: "You don't see anybody waving, do you?"

Borman was a tough commander, more concerned with flying the mission correctly than with its broader poetic significance. At one point he told his crewmates to stop daydreaming and focus on their instrument panels. "I don't want to see you guys looking out the window." When Lovell accidentally inflated the life vest in his survival harness, Borman glared at him. But when Borman wasn't looking, Lovell and Anders took sneaky glances out of the window at the dwindling Earth. They were amazed and disturbed that they could block their entire home planet from view just by holding up a thumb.

Fortunately they also took pictures and brought back home something far more precious than NASA's scientists could ever have imagined. Color photographs of the beautiful Earth rising above the harsh lunar horizon were printed in every major magazine after the mission and humanity began to appreciate, perhaps for the first time, just what a tiny, fragile, lonely planet this is in the vast blackness of space. Bill Anders observed later: "We flew all the way to explore the Moon, and the most important thing we discovered was the Earth."

While America celebrated, a few months later Russia finally recognized it had lost the race to the Moon. Alexei Leonov, one of the cosmonauts who had been training for a possible mission aboard the huge N-1 lunar booster, witnessed the moment of truth in February 1969 when the rocket was test-launched for the first time. "The engines shut down prematurely, and this brought the rocket crashing down to Earth about sixty miles from the launch site. All I saw was a flash in the distance and a fire on the horizon." In July, even as Apollo 11 was undergoing pre-flight checks at the Florida launch complex, Leonov watched another N-1 disaster just a few seconds after it had lifted off the ground. "I saw flames engulf the rocket. It tilted over and collapsed back towards the launchpad with such a deafening explosion that windows in buildings ten to twenty miles away were shattered."

Russia lost the race to the Moon. But as we shall see, it certainly did not lose its space superiority over the years to come.

"I trained hard for many years to fly around the moon and take a close look at it, because I thought that's what the mission was. When we got home, I realized we had discovered something much more precious out there: the Earth. I believe the environmental movement was tremendously inspired by those missions. That alone is worth the relatively few billions of dollars that we spent on Apollo."

Apollo 8 astronaut, Bill Anders

The Language of Apollo

> At Mission Control in Houston, it was 3:17 p.m. when Apollo 11 commander Neil Armstrong at last confirmed the success that everybody had prayed for. "Tranquillity Base here. The Eagle has landed." Television and radio reporters from more than 50 countries were listening to the NASA communications loop and seconds later billions of people around the world were cheering at the news. Everyone was keen to hear what the Apollo astronauts were saying and so the reply from Mission Control CapCom Charlie Duke went almost unnoticed at the time. "Roger, Tranquillity, we copy you on the ground. You got a bunch of guys about to turn blue. We're breathing again. Thanks a lot."

Why had the mission controllers been holding their breaths? Unknown to the general public, Apollo 11's lunar landing had edged close to disaster. The first hint of danger cropped up as the Lunar Module (LM) was plunging towards the surface. The computer display in the cabin suddenly started flashing. "Program alarm," said Armstrong.

Alongside him in the cabin, Buzz Aldrin punched a key on the display board. He reported to Mission Control: "1202." He was calm but slightly puzzled. What the hell was a 1202 code? He couldn't remember running across it during any of the simulations he'd gone through in training.

At Mission Control, 26-year-old Steve Bales was sitting at his guidance console, monitoring telemetry from the LM's computer and navigation systems. Senior flight controller Gene Kranz came on the microphone loop and demanded to know, "What's a 1202?" Bales was in the spotlight suddenly. The LM was rushing towards the moon and the computer was saying something was wrong. He needed a few seconds to think. Above all, he had to keep calm. "Stand by," he replied, trying to buy time. But Aldrin wanted an answer right away. "Give us the reading on the 1202 alarm," he said. In astronaut-speak, he was asking if he and Armstrong should abort the landing. Bales had no time left to think. The LM was plunging into what NASA insiders privately called the "Dead Man's Zone".

After undocking from the orbiting command module to begin its separate voyage, the LM's crew module (ascent stage) could punch away from the lower descent stage at more or less any time if anything went wrong. The LM's guidance system contained a special program to deal with this kind of emergency at a second's notice, even if the main computer's other software was malfunctioning. The ascent stage's separate engine could then power the astronauts back into orbit to rejoin the command module. The same applied on the lunar surface.

The control panel of an Apollo command module was the electro-mechanical nerve center for one of the most complicated machines ever created. Did this triumph of engineering also cause astronauts to behave like machines?

ABORT
MIN
SEC
UPLINK ACTY
TEMP
NO ATT
GIMBAL LOCK
STBY
PROG
KEY REL
RESTART
OPR ERR
TRACKER
VERB
NOUN
VELOCITY
ΔV/RANGE
STBY
BACKUP
VHF RNG
RATE
ATT DB MIN
ATT DB MAX
MAN RATE
LOW
HIGH
MASTER ALARM
ACCEL
G
CMC ATT
IMU
GDC
FDAI
SELECT
SCALE
SOURCE
1/2 CMC
ATT SET
IMU
RATE
HIGH
LOW
TRANS CONTR
PWR
OFF
LIMIT
CYCLE
ATTITUDE
PITCH
YAW
MIN
CMC MODE
AUTO
FREE
SC CONT
CMC
SCS
SPS THRUST
DIRECT ON
DIRECT
MN A/MN B
PWR
ΔV THRUST
NORMAL
LV O2/SPS Pc
ROLL
PITCH

"Balance couple, on. TCA throttle, minumum. Throttle, auto CDR. Prop button, reset. Prop button, okay. Abort stage, reset. Attitude control, three of them to mode control. Okay, mode control is set. AGS is four hundred plus one."

Apollo 11's Buzz Aldrin during the lunar descent, July 1969

Apollo 16 commander John Young (foreground) and his moonwalking colleague Charlie Duke practice working in their bulky suits, while a technician keeps an eye on them, in a scene that suggests the strange fragility of space explorers.

"Everybody at NASA was courteous and saintly, repeating the same information a hundred times, and proud of their ability to serve interchangeably for one another, as if the real secret of their morale was that they had depersonalized themselves."
Norman Mailer, *A Fire on the Moon*, 1970

So long as the ascent engine blast-off was timed to coincide with the orbiting mothership's next pass overhead, the LM's upper stage could get away from the moon pretty sharpish. But the "Dead Man's Zone" was a brief period of uncertainty that nobody could do anything about. Within the last three minutes of the landing approach, there came a time—just ten seconds long—when the LM was hurtling downwards so fast that if its ascent engine fired in an emergency abort-to-orbit, the entire fuel reserve would be wasted counteracting the downward momentum. The ascent stage wouldn't be able to climb back up to join the mothership, orbiting 60 miles overhead. Instead it would crash on the Moon. As the 1202 computer alarm sounded, the Dead Man's Zone was coming up in 20 seconds. If Bales was going to recommend an abort, it had to be now.

According to the code book, 1202 meant that the LM's computer was overloaded and might fail at any second. Making the bravest and possibly the most reckless decision of his life, Bales spoke into his headset for Kranz and all the other controllers to hear. "We're 'Go' on that alarm." This was a signal for the astronauts to ignore their warning lights and carry on with the descent. Kranz was surprised, but he trusted all his controllers completely. He let the decision ride, and the LM continued its vertiginous descent.

Three decades later, Bales is still unsure about the decision he made in that terrifying moment. "Nobody really knew what was causing the problem, and I couldn't be 100 percent certain that the judgment I was making was okay. It was based more on instinct than hard facts."

The next knife-edge judgment call was one for Neil Armstrong. Just as Bales was recovering from the 1202 scare, CapCom Charlie Duke was radioing a terse warning to the LM. "Sixty seconds," is all he said. Listening to the NASA dialogue tapes, it is impossible to tell that anything was wrong, but in technical shorthand, Duke was really saying to Armstrong, "Your descent engine only has sixty seconds' worth of fuel left." In all the simulations that Armstrong had performed while training, he had made it down to the surface with plenty of fuel to spare. What was the problem this time?

Duke called out another warning. "Thirty seconds." Chief astronaut Deke Slayton, standing behind Duke, said softly to him, "Shut up Charlie, and let 'em land."

Any moment now and the lunar module's engine was going to shut down for lack of fuel. The very instant it did so, the ascent stage would automatically separate and begin an emergency return to lunar orbit. Firing the ascent stage under these chaotic circumstances was no guarantee of safety, even out of the Dead Man's Zone, but it was better than simply crashing the entire craft onto the Moon.

Astronaut Rusty Schweikart emerges from the hatch of Apollo 9 for a spacewalk.

Astronaut Rusty Schweikart emerges from the hatch of Apollo 9 for a spacewalk, during an earth-orbital test of the combined spacecraft system.

Kranz and his team then saw something amazing on their telemetry screens. With less than 100 feet remaining before touchdown, the LM was apparently pitching forward and skimming over the lunar terrain at 35 miles per hour. Had Armstrong and Aldrin suddenly lost control of their ship?

Another ten seconds crawled by. At last, Aldrin radioed, "Contact light. Mode control to 'Auto'. Engine Arm off." Those were, in fact, the first words ever spoken by a human being on the surface of another world. But it was his colleague Armstrong who got the credit when he chipped in a moment later with phrases more easily understandable to the general public. "Tranquillity Base here. The Eagle has landed."

Mission Control erupted into applause and Duke's radioed response, "You got a bunch of guys down here about to turn blue," was almost lost among the whoops and hollers of delight.

Armstrong sounded almost apologetic when he radioed to explain what had happened in the last minute of descent. "Houston, that may have seemed like a very long final phase, but the guidance computer was taking us right into a crater with a large number of boulders and rocks." He had needed precious extra seconds to hover about 10 feet above the surface and nudge the ship forward a few hundred yards until he could find a safe place to land. It didn't help that the overloaded computer had steered the lander about 3 miles away from its originally planned touchdown site.

Houston didn't need any apologies. "Be advised, there are lots of smiling faces in this room," Duke replied.

"There are two smiling faces up here also," said Aldrin.

"And don't forget one in the command module," radioed Mike Collins, sitting alone in the orbiting command module.

As so often in big technology projects, the computer problem came down to simple human error. A last-minute decision was made before Apollo 11's launch to switch on two radars during the LM's descent. One would keep tabs on the remaining height above the surface, and the other would beam out in the opposite

Neil Armstrong and Edwin "Buzz" Aldrin after their successful Apollo 11 moonwalk.

direction, tracking the Apollo command ship just in case the LM ascent stage had to make a quick return. The computer was originally programmed to deal with one radar at a time, but the experts thought it should be able to handle the extra load of running both radars at once during the descent. As it turned out, they were almost wrong. The 1202 alarm was a warning to say, "The computer is close to overloading." Close, but not quite. When the time came for the ascent stage to leave the Moon, the computer had only one job to perform: finding the command module again. The 1202 alarm vanished. When Aldrin and Armstrong were safely back in the command module, Collins learned how close Armstrong had come to running out of fuel. "Were you really down to twenty seconds?" he asked. "That's plenty of time," Armstrong replied laconically.

After the mission, Steve Bales collected a certificate from NASA and a medal from President Nixon for "saving" the Apollo 11 mission. He often wondered what they might have given him if he'd been wrong. He also felt compelled to remind people that he was only "the tip of the spear" in making his decision. He had called on help from many backroom software experts within the mission control complex at Houston.

To the general public watching on television, Apollo 11 appeared to be a smooth mission. Even though a few media pundits understood the technical drama quite well at the time, few could convey the frightening significance of NASA's clipped dialogue and shorthand way of talking about the 1202 alarms. The endless repetition, parrot-style, of the NASA jargon soon bored the public, for this was not a language that ordinary people would expect to speak about such supposedly epic events. Soon the events themselves seemed less momentous: just an inevitable product of a Lunar Orbit Insertion and Extra-Vehicular Activity followed at the Time-Nominal moment by a Trans-Earth Injection Burn. Too many journalists were diverted from the drama by this barrier of language, occasionally steeping themselves in it like wannabe astronauts, but more often feeling dislocated by it. As Norman Mailer confessed, "In fact I was bored. Sitting in the Manned Spacecraft Center movie theatre, I noticed the press reporters were also bored. We all knew the engine burn would succeed and Apollo 11 would go into the proper orbit. There seemed no question of failure... I could not forgive the astronauts their resolute avoidance of an heroic posture... They spoke in a language not fit for a computer of events that might yet dislocate eternity."

"If I had a choice, I'm not sure I would want to go again. I went to the Moon. What could I possibly do next? I suffered badly from what the poets have described as 'The melancholy of all things done.' Without a goal, I was like an inert ping-pong ball being bounced about by the whims of other people."

Buzz Aldrin, *Return to Earth*, 1973

Buzz Aldrin descends the ladder of the Apollo 11 lunar module Eagle on July 20, 1969. Although Neil Armstrong was the first man ever to walk on the moon, most of the iconic images from the Apollo 11 mission are of Aldrin, photographed by Armstrong. In this sense, the two men share the glory of that first mission, for if Armstrong is the "name" forever associated with this historic moment, then Aldrin (or at least, his gold-plated helmet) is certainly the "face."

People who fly aircraft or spacecraft tend to keep chatter to a minimum because vital information might be lost if someone takes up too much time talking. NASA's acronyms are like medical terms: precise, exactly defined, highly specialized and almost entirely devoid of emotional loading. A potentially terrifying fuel leak on the launchpad is an "anomalous propellant discharge". A homecoming capsule does not splash into the water right-side up or upside-down, but in positions labeled either "Stable-A" or "Stable-B". A cataclysmic explosion on a shuttle is "a major malfunction". No one dies or is killed. Rather, a crew "fails to respond".

So, why not tell it like it is? NASA is so often criticized for apparently stripping language of its meaning. Mailer was impressed by the professionalism of Apollo 11's mission controllers, and by their calmness and cooperative spirit. Yet he complained that they "spoke in a language not fit for a computer of events that might yet dislocate eternity." Men had ventured to the Moon, but somehow the true majesty of that fact was not reflected by NASA's dry presentation. The dangers of the mission—precisely those dangers that might have excited and moved the general public into understanding just how incredible this achievement really was—were somehow lost in translation, and perhaps even deliberately suppressed.

NASA shares with the military an instinct that potentially terrifying situations can only be handled by keeping some sense of emotional distance. Keeping calm in a crisis is vital for space men and women operating such dangerous machinery. Complicated technical ideas have to be passed from one person to the other swiftly and there is no room for misunderstandings. Astronauts certainly cannot afford the ambiguities of poetry. Their lives often depend on knowing, in a split second, what some particular word or phrase means.

Above, ex-president Lyndon Johnson watches the launch of Apollo 11. Soon the missions would appear deceptively routine, and fewer dignitaries would attend the launches. Opposite, the shot of Apollo 12's crew of Alan Bean, Pete Conrad and Dick Gordon in November 1969 shows three men heading for the moon on a wet Friday morning as if it were the most normal thing in the world.

"This has been far more than three men on a mission to the Moon; more still than the efforts of a government and industry team; more, even, than the efforts of one nation. We feel this stands as a symbol of the insatiable curiosity of all mankind to explore the unknown."

Edwin "Buzz" Aldrin Jr, in a broadcast from Apollo 11

Test engineers occupy a factory-fresh command module inside a vacuum chamber for a three-day session locked inside the ship, checking its seals and life support systems. Opposite, the incredibly fragile ascent stage of Apollo 17's lunar module prepares to rejoin the command module. The skin of the ascent stage was so thin it could be punctured with a screwdriver.

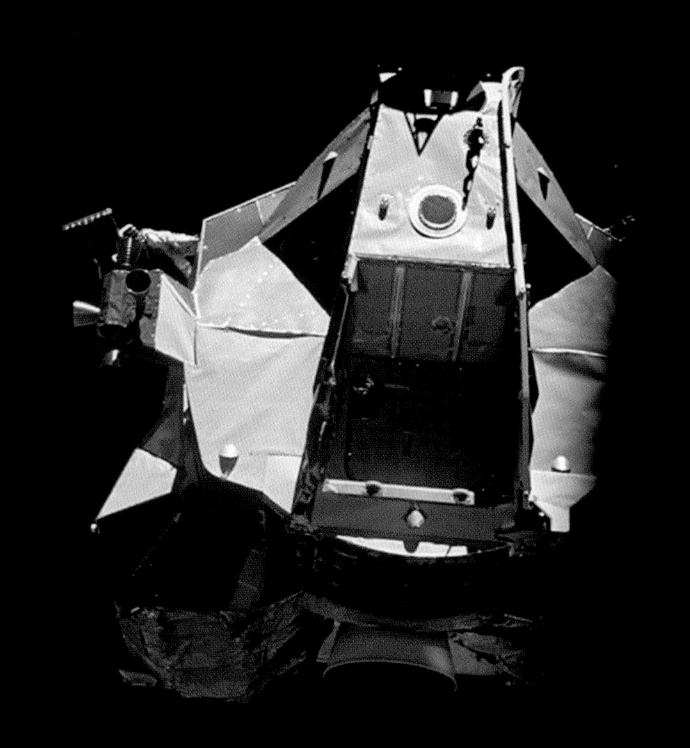

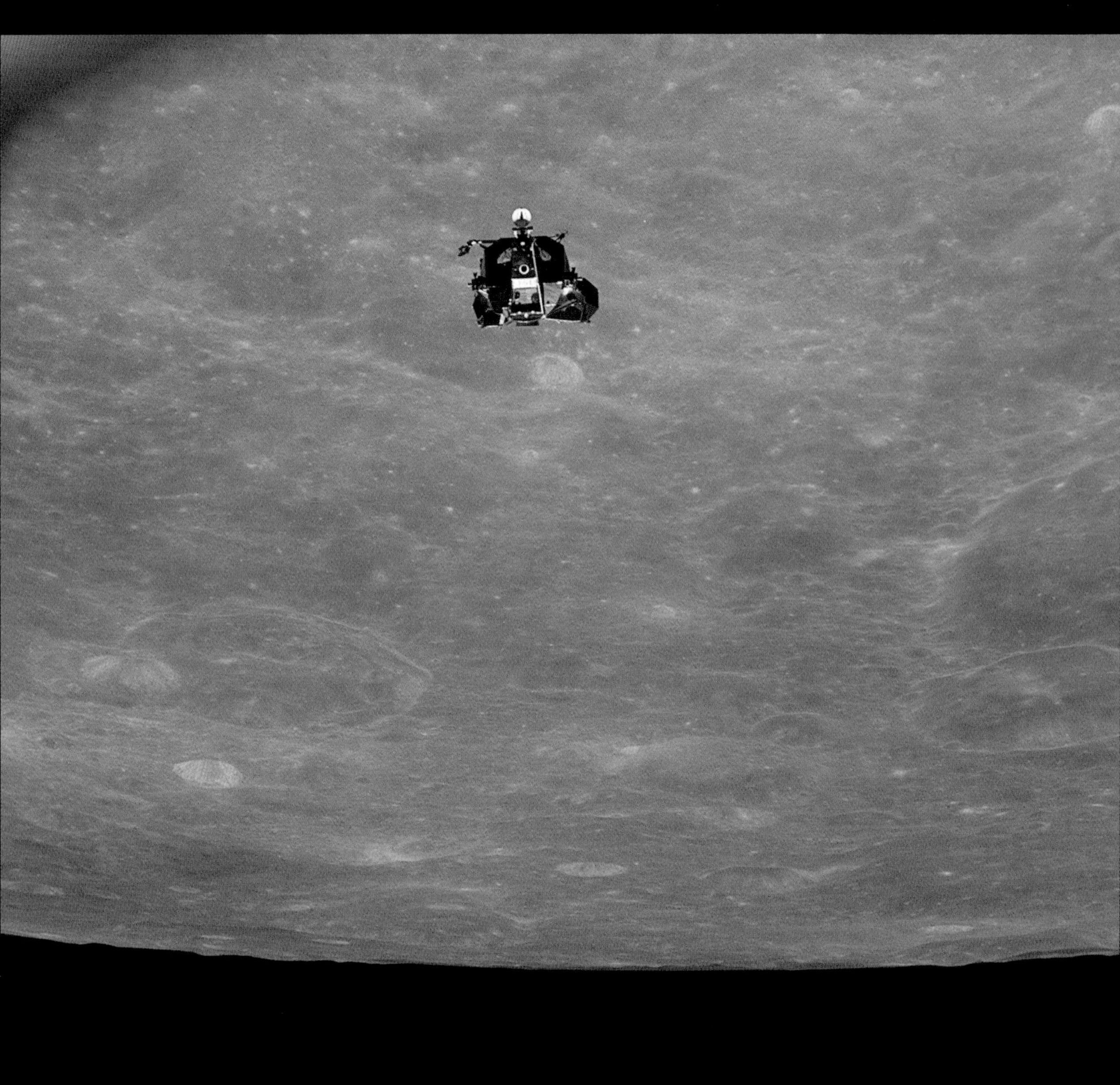

"I think a future flight should include a poet, a priest and a philosopher. We might get a much better idea of what we saw."
Apollo 11 astronaut Michael Collins, 1969

"The most terrifying fact about the universe is not that it is hostile, but that it is indifferent. But if we can come to terms with this indifference and accept the challenges of life within the bounds of death, our existence as a species can have genuine meaning. However vast the darkness, we must supply our own light."
Stanley Kubrick, film director, 1968

"You know what? You're pretty low on fuel. I think you might be in trouble. You sure you guys understand what you're doing? According to my readings here at Mission Control, you've only got about a minute left," is not so useful to an astronaut as the quick, simple reminder: "Sixty seconds."

This unavoidable gulf between NASA-speak and the rest of the world remains a problem today, for people are still surprised when missions go wrong.

No language has yet been invented that can persuade the public to fund a basically very dangerous project. They expect NASA to anticipate every risk, and that somehow space vehicles fuelled with the energy equivalent of small thermonuclear bombs can be made as safe to fly as ordinary aircraft—at least, they should be if the government is paying for them with tax dollars. Surely all that dry technical babble must mean that NASA understands what it's doing by now, and that failures must therefore be the result of incompetence? But rockets are dangerous and outer space will always be a hazardous environment. People make errors of judgment, and they can be very serious ones that lead to disaster. And equally, sometimes people's animal instincts save a mission even when all the technical dials and read-outs are indicating trouble. Apollo 11 could so easily have crashed, plunging the United States into an international humiliation that would have echoed to this day. Or the shuttle Columbia could so easily *not* have been struck by fuel tank debris in January 2003.

The danger, the fragility and the courage of space exploration is conveyed by this image of a lunar lander ascent stage heading back into orbit.

Humans are not angels. The machines of spaceflight are the imperfect products of imperfect people working as best they can within complicated organizations. Yet we reach into the heavens despite our flaws and this is what gives the space adventure its true grandeur. It's not the fact that we explore space fitfully and uneasily, so much as that we manage to do it at all that counts in our favor. Our continued sallies into space in the coming generation will remain, as John F. Kennedy told us back in 1961, "difficult" and "expensive to accomplish".

To date just under five hundred people have flown into space: some just once, many repeatedly. Of those five hundred, three were killed before their craft ever left the Earth, another crashed to the ground, three were suffocated in space, seven were blown apart soon after launch, and seven more died in a high-altitude disintegration of their ship. This, then, is a fatality rate of one space traveler out of every twenty-three. A dangerous profession indeed. Yet somehow the illusion persists that space travel has become routine and that accidents are unusual—and blameworthy.

Hundreds of thousands of people are fascinated by space exploration and millions more take at least a

This photo of Alan Bean during the Apollo 12 moonwalk was taken by his colleague Pete Conrad more than three decades ago, yet it seems timelessly fresh.

passing interest, but those multitudes are perhaps not sufficient to ensure a long-term continuation of the human space program at a major national level. The drama of human rocket flight is as great now as it ever was. Yet astronaut-heroes do not seem to captivate whole societies as they once could, and rockets have lost some of their ability to inspire awe.

There is a problem in the language of space if young people no longer find as much excitement in the adventure as their parents did fifty years ago, when they, in their turn, were young. Perhaps NASA and its international partners need, after all, to speak to this generation in a more poetic way that engages emotions as well as minds.

In October 1960, a new prototype Soviet rocket, the R-16, was hoisted upright for launch at Baikonur in Kazakhstan. This was supposed to be a fast-response military strike alternative to Korolev's R-7, with space capability on the side, but instead it created the single greatest rocket disaster in history.

As so often with rocket tragedies, politics played a part. That same October, Soviet leader Nikita Khrushchev had just come home from a United Nations conference in New York, where he'd bragged about Soviet military superiority, with the secret R-16 very much on his mind. "We're turning out missiles like sausages from a machine!" he shouted. Now he wanted proof. He ordered his chief of ballistic missile deployment, Marshal Mitrofan Nedelin, to get down to Baikonur and supervise the R-16's debut flight.

OFF
MAIN MOM
TALK

"My interview with Apollo 16 moonwalker Charlie Duke was just ending when his wife Dotty came into the room after taking a phone call. She had shocking news. Apollo 12 astronaut Pete Conrad was dead. Duke's eyes clouded over as he talked about his comrade. Something he said to me sent my mind reeling. 'Now there's only nine of us left.'"

Andrew Smith, *Moondust*, 2005

Lunar module pilot Charlie Duke explores the stark lunar landscape of Plum Crater during the Apollo 16 mission of April 1972.

"It looks to me, looking out the hatch, that we are venting something out into space. It's a gas of some sort."
Apollo 13 commander Jim Lovell, April 13, 1970

In theory, the R-16 could be fuelled and primed several days, or even weeks, before it was needed, because its nitric acid and hydrazine fuels could be held for long periods inside the rocket at normal pressures and temperatures. The trouble was, these "storable" fuels wouldn't store. They were viciously corrosive and they leaked. The R-16 needed much more attention than its designer (Mikhail Yangel, a Korolev rival) had promised. On October 23, the surrounding launch gantries were crowded with young technicians trying to fix problems. An incorrect command was transmitted to the R-16's upper stage. Its engine fired, straightaway burning a hole in the top of the stage beneath it. This lower stage exploded, instantly killing everyone on the gantry. With nothing to support it, the upper stage crashed to the ground, spilling fuel and flames. The new tarmac aprons and roadways around the gantry melted in the heat, then caught fire. Ground staff fleeing for their lives were trapped in the viscous tar as it burned all around them. The conflagration spread for thousands of yards, a wave of fire engulfing everything and everyone in its path. More than 190 people were killed, including Nedelin, perched on his chair near the gantry as a surge of blazing chemicals swept towards him.

For 30 years, the West knew little of this, although it was apparent from intelligence reports that something had gone wrong. All of Soviet Russia was saddened to hear (eventually) that Marshal Nedelin and several "senior missile officers" had been killed in an "aircraft accident". Of course the absence of 190 familiar faces became obvious to thousands of space workers beyond Baikonur, but nobody (not even the many grieving mothers) was allowed to discuss the tragedy.

More recently, accidents in China, India, Japan and Brazil have claimed dozens of lives, and shown that rockets will always be exceptionally hazardous machines to work around.

Apollo 13's command module is recovered after splashdown, at the climax of an incredible adventure in April 1970. The rear service module had exploded on the way to the moon. Astronauts and mission controllers devised between them an incredible rescue, in which the lunar module was transformed from a landing craft into a lifeboat, carrying the crippled command module back to earth.

"Nothing is more symptomatic of the enervation, of the decompression of the Western imagination, than our inability to respond to the landings on the moon. Not a single great poem, picture or metaphor has come of this breathtaking act."
US critic and novelist George Steiner, 1994

"This is the greatest week in the history of the world since the Creation."
President Richard M. Nixon, July 1969

"The Lunar landing of the astronauts is more than a step in history. It is a step in evolution."
The New York Times, July 20, 1969

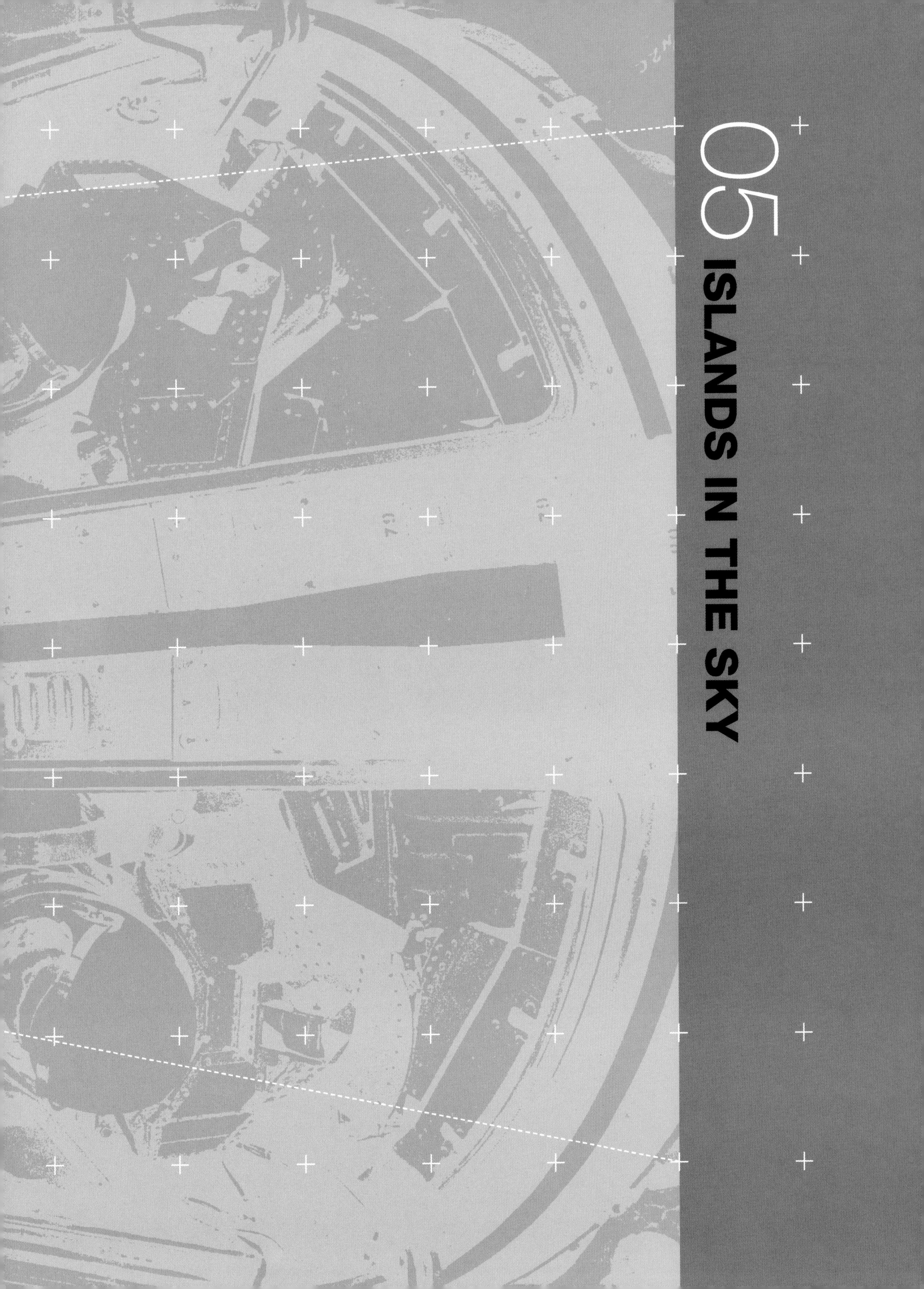

05 ISLANDS IN THE SKY

"Scientists and engineers now know how to build a station in space that would circle the earth 1,075 miles up. The job would take 10 years, and cost twice as much as the atom bomb. If we do it, we can preserve the peace and take a long step toward uniting mankind."
Collier's Magazine, 1952

ISLANDS IN THE SKY

> The basic idea for a space station emerged more than a century ago in *The Brick Moon* (1869), a story by the American author Edward Everett Hale. He describes a brick sphere 200 feet in diameter, thrown into orbit by a pair of colossal flywheels. The sphere is intended to be a beacon, a reflective blip in the sky that can be observed by ocean-going ships and used as a reference for navigation and timekeeping. But some of the sphere's builders accidentally become passengers when it is hurled aloft prematurely. They communicate their plight using mirror semaphore. Back on Earth, the flywheels are powered up again to send food and supplies up to the stranded workers, and so a strange accidental Space Age begins.

By the 1920s, real-life astronautical theory was advancing at a great pace. Rocket pioneers such as Konstantin Tsiolkovsky in Russia, Robert Goddard in America, and Hermann Oberth and Wernher von Braun in Germany understood the potential for space stations. The first genuinely detailed engineering proposal appears to have been prepared by Hermann Noordung, an engineer in the Austrian army. His short book *The Problem of Space Travel* (1927) influenced rocket engineers for the next half-century.

Very little of Noordung's life story is known today, except that he died tragically young of tuberculosis just two years after publishing his brilliant ideas. Fortunately his space station proposal survives in all its prescient detail. He conceived a "habitation wheel" whose gentle rotation provides its crew with artificial gravity. Airlocks and safety bulkheads are all described, along with a huge parabolic dish that collects and focuses sunlight for power. The docking airlock rotates at the same rate

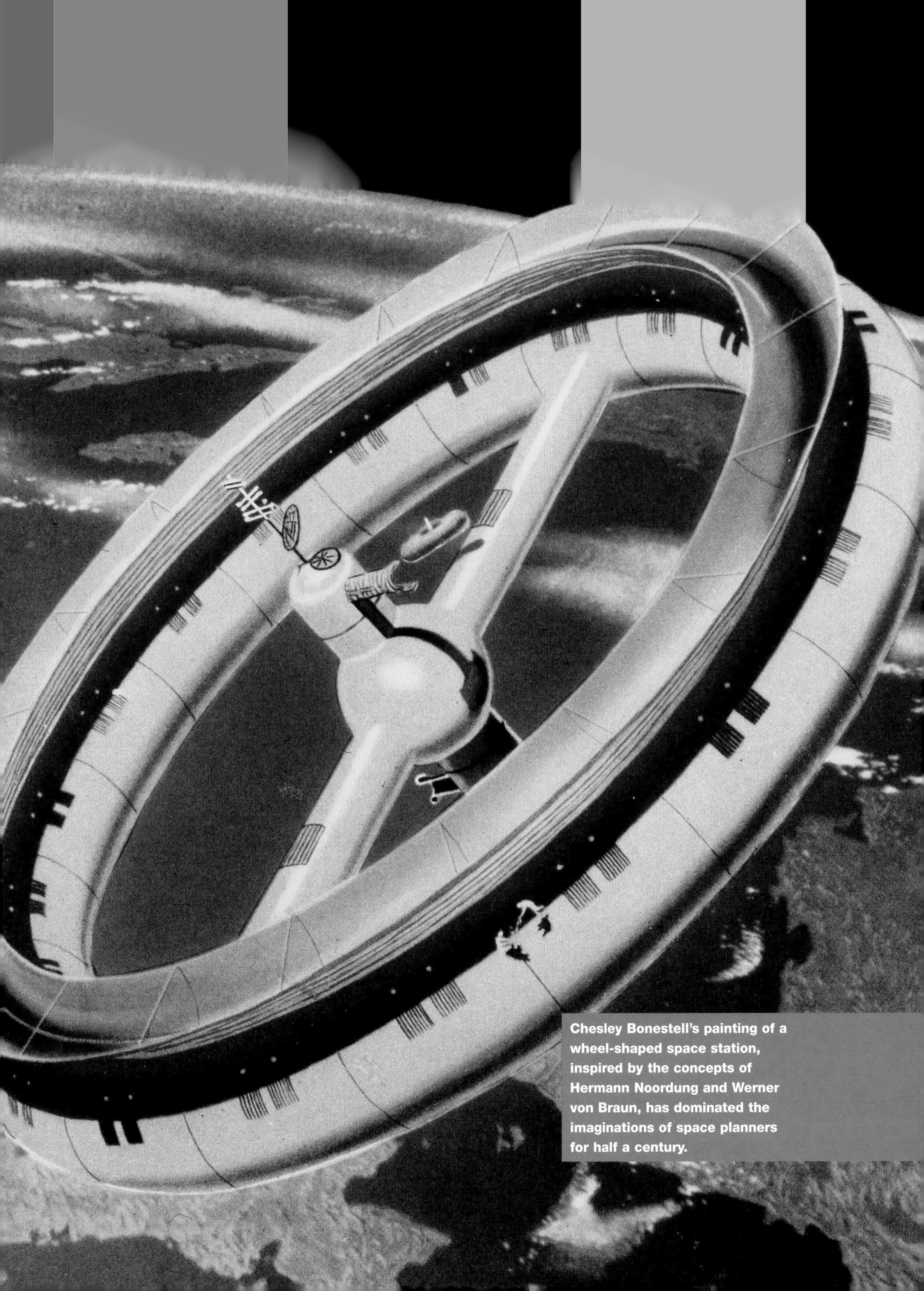

Chesley Bonestell's painting of a wheel-shaped space station, inspired by the concepts of Hermann Noordung and Werner von Braun, has dominated the imaginations of space planners for half a century.

as the station, but in the opposite direction, maintaining its position relative to the horizon. This allows rocket ships to approach without themselves having to tumble. But what practical use might a station serve? In the early 1940s, a young British radio expert and budding science fiction writer, Arthur C. Clarke, helped to develop navigation systems for allied aircraft returning from their bombing raids over Germany. At the end of the war he recognized that his dream of spaceflight would not be realized unless governments and industrial investors could be persuaded to see its economic benefits. In October 1945 he published an article in an electronics magazine, *Wireless World*, essentially outlining the modern concept of geostationary communications satellites. The only difference between Clarke's ideas of 1945 and the global network that we see today is that he was thinking in terms of bulky and unreliable 1940s radio equipment. He suggested that his "extraterrestrial relays" would have to be permanently occupied space stations tended by human engineers. Clarke was a leading member of the British Inter-planetary Society. Founded in 1933, this amateur but very disciplined organization became an influential study group in the field of astronautics. In 1939 it demonstrated that technology available in that generation could be used

to send a rocket to the moon, if only someone would spend the money. By 1946 Clarke's colleagues Harry Ross and Ralph Smith had devised a rotating space station similar to Noordung's proposal: a detailed version of the platform envisioned by Clarke as a communications relay. It came complete with a radio room, library, kitchen, surgery, small cabins for the crew, a slightly better office for the Chief Engineer and the best quarters for the Station Director. In many ways it was a very English kind of space station.

When von Braun adopted the space station as part of his *Collier's* manifesto in the 1950s, he thought of it as a staging post between the Earth and other worlds. Astronauts housed in it would build gigantic interplanetary spaceships using components launched on smaller rockets and space planes. But a decade later, NASA engineers conceived and built Apollo's ultra-lightweight lunar landing module, eliminating any requirement to assemble giant, heavy ships in Earth orbit. Each Apollo mission required only a single launch, and the idea of a space station as a hotel for construction workers became redundant almost as soon as the Space Age began. Just as significantly, the electronics and other complex systems inside the Apollo were miniaturized to a degree that the prophets of space could scarcely have imagined back in the 1950s.

And yet the dream refused to die. In 1968, just as Apollo was about to reach the moon, Stanley Kubrick's movie *2001: A Space Odyssey* (co-scripted by Arthur C. Clarke) delivered perhaps the most persuasive space station in all of science fiction: a gigantic "Orbiter Hilton" with garishly colored designer chairs in the lounge areas, phone booths and coffee vending machines on standby, and panoramic windows offering spectacular views of the Earth from the vantage point of high orbit—a view so familiar to the spacefarers of *2001* that they practically ignored it.

Fred Freeman's artwork for *Collier's*, above, shows a cutaway of a large rotating space station. A decade later, the movie *2001: A Space Odyssey* predicted that travel to orbit would become a routine experience opposite.

The First Real Stations

> Throughout the late 1960s the Soviet Union attempted a moon landing project of its own, in conditions of absolute secrecy. As we have seen, the gigantic N-1 lunar booster was never fit to fly. The Soviets fell back on using smaller but more reliable rockets to exploit the possibilities of Earth orbit. To them goes the honor of having built and launched the world's first manned space station.

On April 19, 1971, Salyut (Salute) was launched atop a Proton booster. It was a cramped, stuffy, cylindrical compartment barely 6 feet in diameter at its widest point and 50 feet long. The interior space compared unfavorably with a Greyhound bus filled with luggage, but Salyut provided more living room than any space traveler had enjoyed before. Three days later a Soyuz crew ferry made a rendezvous with it, although the trio of cosmonauts inside the cramped capsule did not transfer to the station. Their docking system failed to achieve a proper seal.

On June 6, cosmonauts Georgi Dobrovolsky, Vladislav Volkov and Viktor Patsayev lifted off in another Soyuz to become the first occupants of a home in space, living aboard Salyut for 23 days. As it turned out, those days were precious indeed.

Since Vladimir Komarov's accident in 1967, further test flights of the Soyuz seemed to have improved the system, but when Salyut 1's crew unlatched from the station to begin their reentry, all the air leaked out of their cabin through a faulty valve. The return capsule came smoothly back down to Earth, but Dobrovolsky,

Above, the unfortunate crew of Salyut 1. Opposite, ground staff inspect the base of a Soyuz booster rocket. Supercold fuel in the tanks freezes moisture in the surrounding air, causing frost to form on the rocket's skin.

"We were flying over America and suddenly I saw snow, the first snow we ever saw from orbit. I have never visited America, but I imagined that the arrival of autumn and winter is the same there as in other places, and the process of getting ready for them is the same. And then it struck me that we are all children of our earth."
Salyut 7 cosmonaut Aleksandr Aleksandrov, 1983

Volkov and Patsayev were already dead by the time the parachutes had unfurled. Recovery crews were appalled at how deceptively normal the cosmonaut's bodies looked when the capsule's hatch was opened. It was not possible to despatch another crew to Salyut 1. Six months after its launch, the station's orbit had decayed and it fell into the atmosphere, burning up in a remote-controlled suicide assisted by ground controllers. The Soyuz capsules were redesigned to higher standards (in fact the design is so reliable that they are still in use today) and six more Salyut stations were flown between 1972 and 1982. The largest item of equipment on board several of the stations was a high-resolution camera through which cosmonauts dutifully peered down at NATO exercises and other Western military activities. Official Soviet press reports mentioned "Earth observations in the interests of the national economy".

Genuine science was not forgotten, however. The Russians gained substantial experience in astronautical medicine, biology, astronomy, metallurgy and the myriad activities that make up an active life in space. From 1975 they learned how to resupply their Salyuts with unmanned robot ships, Progress ferries loaded with oxygen, food and water. Progress ferries were adapted from Soyuz hardware, with the crew cabins stripped out to make room for cargo. It was an efficient piece of design, building on what was already available.

In February 1985 the final Salyut station had been unoccupied for a few weeks and another crew was scheduled to fly up and join it. But the station's solar panels had lost their orientation to the Sun. With the navigation systems starved of electrical power, Salyut 7 gradually drifted out of control. Meanwhile, the air-conditioning systems failed and the interior cabin space filled with toxic fumes. Down on the ground, cosmonauts began training for a rescue mission. On June 16, Vladimir Dzhanibekov and Viktor Savinykh warily piloted their Soyuz capsule around the flailing station like boxers circling just beyond the reach of an angry opponent. They worked out a way of docking and then made repairs.

This last Salyut hosted guest astronauts from other Communist countries, and even from France and India. Soviet envoys happily offered flight time on board their station to any Americans who might want it. NASA officials politely declined through gritted teeth and prayed that the agency could get its own space station developed soon. A second space station, that is. The first one was called Skylab. The general public scarcely remembers it these days, yet it was the largest spacecraft ever launched on a single rocket.

The iconic movie poster artwork for *2001: A Space Odyssey* depicts a habitat in space for hundreds of people. Reality has yet to catch up.

Skylab

"I could give you guys some lessons. The people who designed this little beauty should be put in front of a firing squad."
Skylab astronaut Ed Gibson, 1973

> When the Apollo lunar landings came to an end, NASA had intended the Apollo Applications Program to take over. It would use the vast power of the Saturn V rocket to launch an ambitious series of space stations into orbit. Budget cuts forced mission planners to scale down the program, but one project survived. The Skylab Orbital Workshop, launched in 1973, still holds the record as the largest spacecraft ever launched on a single rocket and provided American astronauts with their first taste of long-duration orbital flight.

The big factor in Skylab's favor was its economy. It was already half-built even before it was formally approved as a new project in 1969. The third, uppermost stage of a Saturn V rocket was adapted as the station's main compartment. This stage was supposed to carry fuel for an Apollo spacecraft's trip out of Earth orbit and to the moon, but NASA realized that if this stage remained in Earth orbit, the propellants and rocket engines for the moon journey would not be needed, and the huge fuel tanks could be filled with air instead of fuel. They could be split into cabins and work areas using lightweight floor and wall dividers.

In 1969, up-coming Apollo moon landing missions 18, 19 and 20 were cancelled, and a spare Saturn V became available, along with three Apollo command modules to serve as crew ferries. Skylab was completed in 36 months for a budget of $2.5 billion, and the station was blasted aloft on May 14, 1973, on the last Saturn V booster to fly.

The lift-off appeared to be flawless and Skylab climbed to a perfect orbit, circling earth at an altitude of 270 miles. But it later transpired that shielding on the lab's outer skin designed to protect against micro-meteoroids and solar heating had been torn off by aerodynamic stresses during the flight. One of the two solar panels was jammed shut in its launch position and the second panel was completely torn away, like a bird's wing ripped off at the shoulder. Skylab's electrical systems were powerless without them. There was no choice but to postpone the launch of the first three-man crew on a separate, smaller rocket (a Saturn 1B) by 10 days. Astronauts Joe Kerwin and Paul Weitz, along with their commander, Pete Conrad, began training for an emergency spacewalk, testing out various repairs for Skylab in a giant water tank, using a full-scale mock-up of the station.

On May 25, Conrad and his crew docked with Skylab, although the nose of their command module ferry had initially seemed reluctant to mate properly with the station. Then they began their extraordinary spacewalk. They freed the surviving solar panel so that it could open up properly and then deployed a shield of metal foil to protect the lab's main compartment against the Sun's heat. They had to push the shade, folded in a tight bundle, through the airlock of the command module and then unwrap it and tie it down against the station using thin straps. This was no job for beginners—but NASA had never tried anything like it before. When the astronauts went inside Skylab they found it had become a giant oven, with internal temperatures of 125 degrees Fahrenheit. They had to

Skylab in orbit, seen from an approaching Apollo crew ferry. One of its solar panels is deployed, and the other has been torn away by aerodynamic stresses during launch.

spend the first couple of days in the cramped docking module at the front end of the station until the solar shade had cooled the sweltering main workshop area and they could gradually bring the life-support systems on-line. But nothing could dampen their mood. Conrad and his crew had saved Skylab from disaster in a dramatic rescue that made headline news.

Two other crews visited the station over the next year. As well as conducting intensive medical experiments,

Rebellious Skylab crewmen Jerry Carr, left and Bill Pogue in a rare off-duty moment.

they operated a sophisticated solar observatory that included X-ray, infrared and visible light cameras, yielding the most detailed observations of our Sun that have ever been recorded.

And that is more or less the approved NASA version of how America's first space station was saved and occupied by brave astronauts. As so often with official accounts of spaceflight, however, a few background problems were not publicized at the time. The last of Skylab's three-man teams had come close to mutiny and changed forever the relationship between managers on the ground and astronauts in orbit.

Ed Gibson, Bill Pogue and commander Jerry Carr were the third and final Skylab crew. When they clambered on board in November 1973 for a 12-week tour of duty, expectations were running high. The previous crews had worked well, setting an enthusiastic pace and getting along fine with Mission Control. But Carr's team got off on the wrong foot. Pogue took a while adjusting to weightlessness and threw up almost as soon as he entered the station's huge workshop module. Not wanting to embarrass him, Carr said: "We won't mention that. We'll just throw the mess down the trash airlock." Pogue agreed: "Nobody will keep track. It's just between you and me."

Imagine their surprise when, the next day, Mission Control gave them a severe reprimand for not reporting Pogue's sickness. As it turned out, their private conversations were being monitored. Everything they said was typed up and copies were on the desks of senior NASA managers within 24 hours. Skylab was bugged like a space version of the Nixon White House. This invasion of privacy really spoiled the mood and several weeks of tension ensued between Mission Control and the orbiting crew.

At first, Carr's team took out their frustrations on Skylab itself. "We really need a better sense of up and down with a proper difference between the floor and ceiling," Pogue complained on the radio link with Mission Control. "The layout in the docking adapter is so lousy I don't even want to talk about it." Engineers had mounted equipment, control panels and storage lockers on floors, ceilings and walls, making use of every available area of interior space. As a result the crew's sense of distance became warped because the horizon that guides us in our daily lives was not apparent in some of their enclosed work areas. In addition, the usual differences in light levels—brighter above our heads, darker at our feet—were cancelled aboard Skylab by the poor arrangement of light fittings, which were sometimes mounted on the ceiling, sometimes on the walls. The astronauts hardly knew from one moment to the next which way round they were facing.

Next on their list of grievances was the poor arrangement of Skylab's storage lockers. They could never find anything. When they did work out the right locker, everything inside it would tumble out as soon as the door was opened, and then all the bits and pieces that they didn't want—pens, screwdrivers, cloths, notepads, film cartridges—would have to be retrieved and stuffed back inside. Soon the entire station filled with drifting junk and the astronauts were in danger of breathing in and choking on a lost screw, perhaps, or a stray shred of paper. The best bet was to wait for a few hours until all the loose debris had been caught by the gentle suction of the air-conditioning filters, which then had to be cleaned out by carefully removing dust and small flecks of rubbish from the grilles with a vacuum hose. This would lead to arguments with Mission Control because the managers down there hadn't scheduled this unforeseen but important task into the astronauts' already busy day.

Then the station's decor started to get on the crew's nerves. "The color scheme in here has been designed

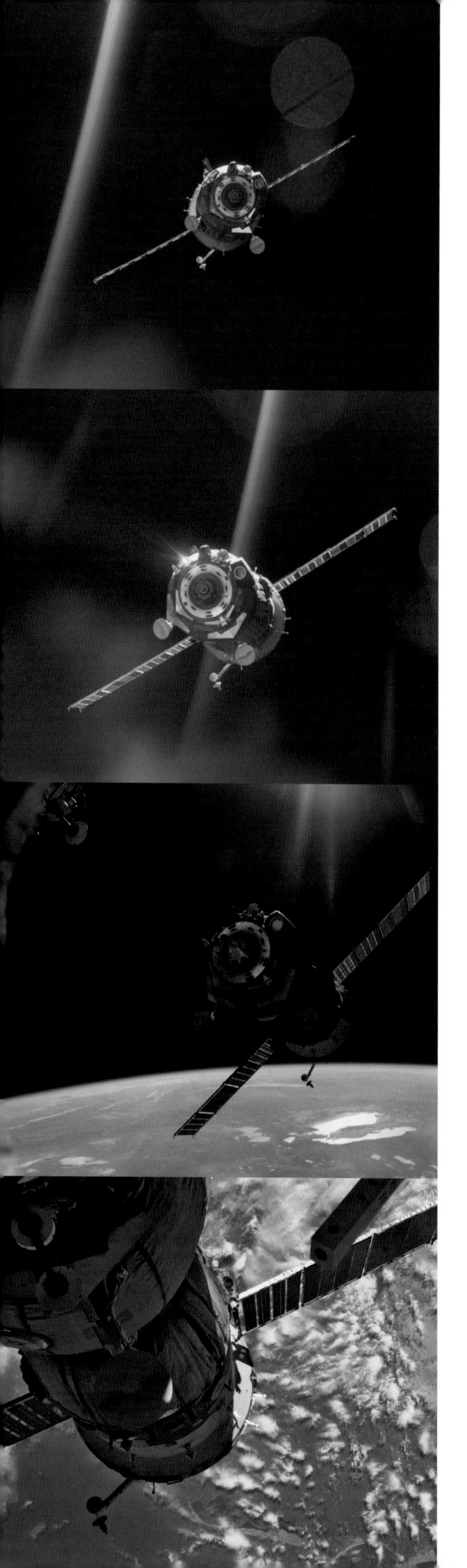

The simple but reliable Russian Soyuz hardware serves as crew transporter, robot-controlled cargo ferry and standby rescue craft for the Space Station.

with no imagination," Gibson complained. "All we've got is about two shades of brown, and that's it for the whole lousy spacecraft interior." The crew's carefully fireproofed garments were as drab as the wall colors, and the artificial fabrics were stiff and prickly against the skin. They all yearned for floppy tee-shirts and casual, comfortable gear in cheerful colors.

There was more to worry about than bad fashion days. Skylab's bathroom was unpleasant to use. The metal floor was designed as a hygienic wipeable surface. It had no firm footholds and at wash times the astronauts slithered about like sardines in a can. Having a good wash took up not minutes but hours, because every last droplet of water that strayed outside the plastic shower enclosure had to be mopped up to prevent moisture seeping into electrical equipment and short-circuiting it. This extra housework cut into the few precious moments set aside for the astronauts to relax. When Mission Control tried to discuss this problem, Carr was too wound-up to listen. "Off-duty activities? You gotta be kidding! There's no such thing up here. On a day off, the only difference is you have time to clean the shower!"

This was the real problem. Carr's team was working long hours on Skylab's many experiments, but somehow they weren't getting much done. They were frustrated and worn-out. Ground controllers relayed endless requests from the scientists who had designed the experiments, but the astronauts struggled to put those requests into action. Skylab's senior flight director at mission control, Neil Hutchinson, couldn't

"For the Russians, space is still an important icon, despite their economic problems. Space has a mystical significance for them. Every year they celebrate Yuri Gagarin and Sputnik. They know that these are everlasting achievements in the great human story that can never be repeated. There's got to be a deep-seated passion there. For privileged Americans with multi-billion dollar budgets, space is just another shop in the mall."
NASA astronaut Story Musgrave, 1996

see the astronauts' point of view. The more they complained, the more Hutchinson piled on the pressure, like an army drill sergeant trying to bully his men into obedience.

Most depressing of all, the crew's tight work schedule prevented them from enjoying the actual experience of being in space. Every spare moment, they huddled around Skylab's single small window staring in childlike fascination at their home world, seeing it as very few people before or since have ever had a chance to see it. But there weren't any spare moments. NASA had told the astronauts that Skylab was about "learning to live in orbit". If this was "living" then it wasn't up to much. One day towards the end of the sixth week, Pogue was struggling with a scientific package and exploded into anger. "There's no way we can do a professional job. I don't like being put in a position where I'm taking somebody's expensive equipment and thrashing about wildly, trying to get it started in insufficient time!" Then Gibson chipped in with a sour complaint about another experiment he was trying to operate, again under relentless time pressure from the ground. "I could give you guys some lessons. The people who designed this little beauty should be put in front of a firing squad."

Carr finally called a strike. "I get the feeling that there's some hassle about who on the ground gets our time and how much of it. We hoped you would have got the message that we did not plan to operate at this pace. We need time for relaxation." For an entire day, the three men did nothing but stare out of the window, take photographs and catch up on personal tasks. Meanwhile, flight director Hutchinson had a rethink and decided, to nobody's great surprise, that maybe he was pushing the astronauts too hard.

Keeping in mind that this was still a new experience for everybody at NASA, the occasional complaints were seen as valuable learning tools, both then and now. In fact, throughout the design process for the International Space Station (ISS), Carr and Pogue served as consultants to NASA and its main contractor, the Boeing Corporation. In advanced middle age, they were still trim and fit enough to clamber into spacesuits and practice procedures in that all-too familiar giant water tank. After more than a quarter of a century, the difficulties of life on Skylab remained fresh in their minds. They claimed a special field as their own: how *not* to design a space station.

Today, NASA is keen to avoid making the same mistakes. The astronauts who live and work aboard ISS are consulted about every last aspect, from the details of spacewalk procedures to their choice of soft drinks. But no matter how many astronauts offer their opinions, much comes down to personal taste. In 1991, someone decided that pink was a restful color suitable for the ISS living quarters. Shuttle pilot Joe Allen wasn't so sure. "I don't live in a pink house. Why would I want to live in a pink space station?"

Opposite, Mir in orbit, photographed from an approaching US shuttle.

The End of Saturn

> On completion of the Skylab progam there were no more payloads authorized for von Braun's great Saturn V launch vehicle. Historians and economists will no doubt argue for years how America managed so quickly to abandon a rocket of such proven power and reliability in favor of a new and untested concept, the space shuttle. At the time it was felt that the huge ground facilities and thousands of staff required to fly the Saturn were no longer justifiable. In the event, the shuttle proved no less expensive to fly. As the Apollo-Saturn program drew to a close, the Saturn's great architect, Wernher von Braun, found himself sidelined to a privileged but largely powerless office in NASA's Washington headquarters, from where, with admirable good grace, he contributed to the political "selling" of the shuttle concept. He retired from NASA in 1972 and died of cancer five years later, with all too many of his dreams for the exploration of the solar system—and especially Mars—left unfulfilled. Russia and America now began to realize at least some of his *Collier's* visions of giant space stations in earth orbit. Yet both countries forgot one important element. A space station was supposed to be just a stopping-off point on the way to other and more ambitious targets. Real-life stations were nothing of the kind. They were an end in themselves, and in their sometimes dullgo-nowhere capacity, they dimmed the public's remaining interest in human space flight.

Space Station Mir

> The Soviets launched the first component of their new-generation Mir (Peace) station in February 1986. The following month, Mir's first crew, Leonid Kizim and Vladimir Solovyov, blasted off for a rendezvous and then clocked up an impressive range of activities. After two months aboard Mir they took their Soyuz capsule for a 50-day visit to an old station, Salyut 7, which was still in orbit. They made two spacewalks, investigating the exterior condition of Salyut and practicing assembly work in space. Then they flew back to Mir, reoccupying the new station for three weeks before finally heading home after a mission totaling 125 days. By any measure, the Soviets had consolidated their superiority in earth orbital operations, using relatively inexpensive, standardized technology to achieve feats that NASA had little immediate prospect of matching.

In the course of their missions, Russian cosmonauts laid claim to significant records: the longest-functioning crewed space vehicles; the greatest distances traveled in orbit; the greatest number of crew changes; the greatest total accumulated mission time in space, and the longest periods spent aloft by individual humans. Even as the old Soviet order passed into history and government-funded space activities were cut to the bone, it remained politically acceptable for Russian space workers to plan for the impossible.

A closer view of Mir reveals its complex multi-module construction and adaptable multiple docking ports.

esa

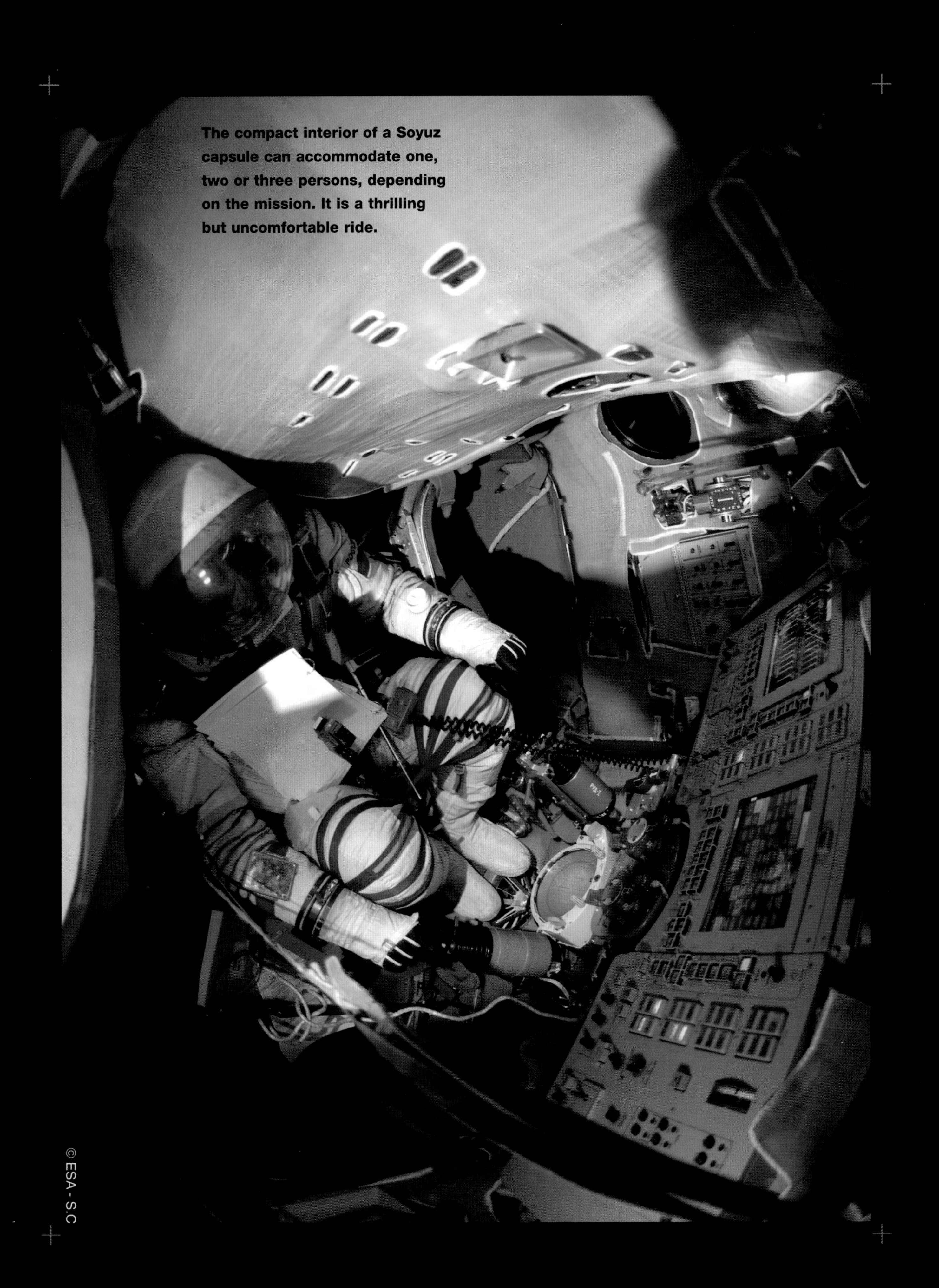
The compact interior of a Soyuz capsule can accommodate one, two or three persons, depending on the mission. It is a thrilling but uncomfortable ride.

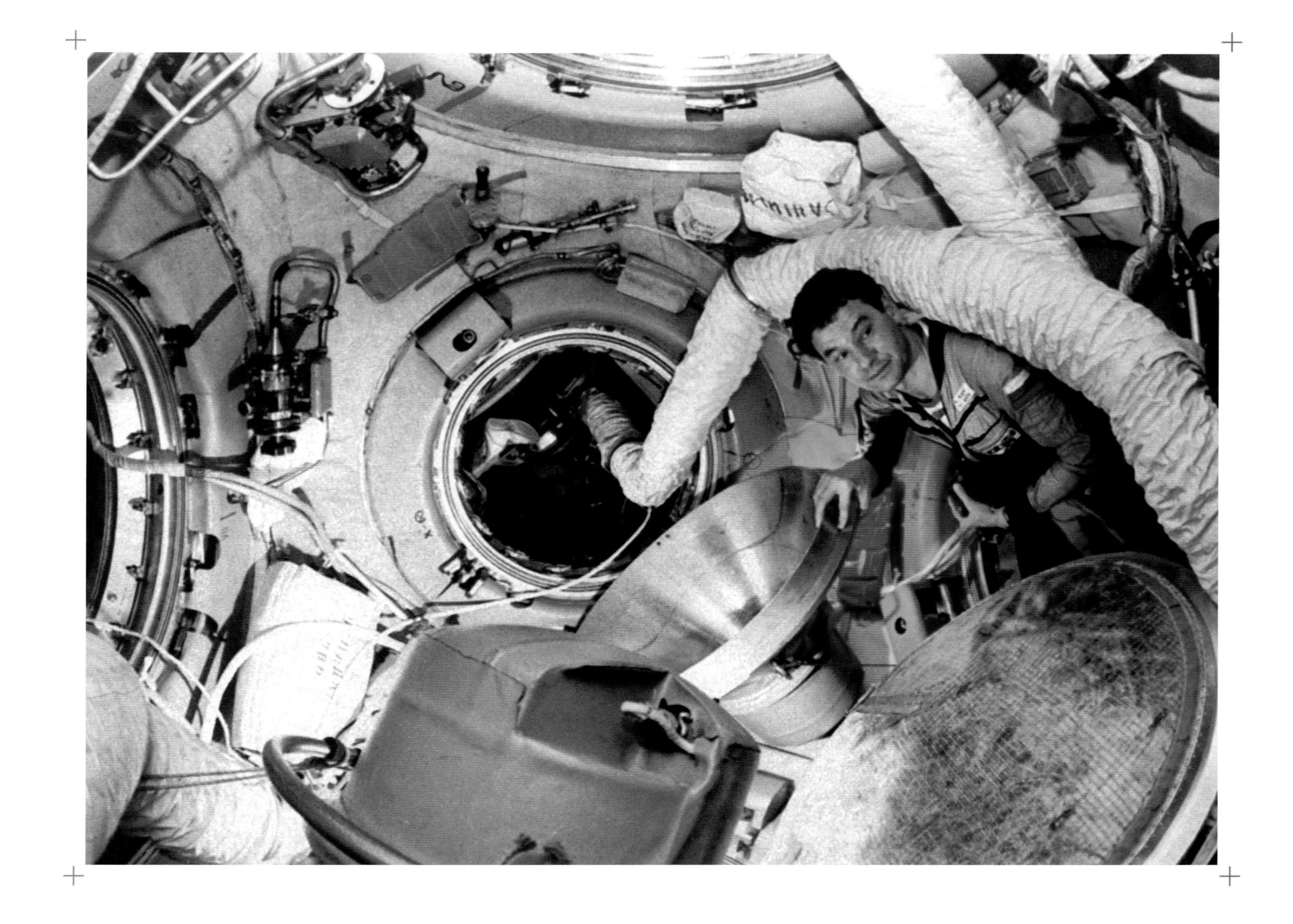

"The Soviet press was obsessed with everything that happened in orbit, including elaborate descriptions of the cosmonaut's menu at their last breakfast, and all the details of their physical exercise program... The relentless propaganda spoiled the very idea of man's flying into space."
Roald Z. Sagdeev, *The Making of a Soviet Scientist*, 1994

They were attempting to gain experience for a flight to Mars. Their cosmonauts wanted to prove that they could survive mentally and physically inside cramped, weightless cabins on such an immense journey. It seemed a curiously optimistic and touching ambition to conquer another planet when Russia faced so many difficulties on earth. As NASA shuttle astronaut Story Musgrave has lamented, "For the Russians, space is still an important icon of national culture, despite all their economic problems. For us, it's just another shop in the mall."

Seven American astronauts visited Mir for prolonged tours of duty between 1995 and 1997, with NASA taking the opportunity to practice docking its winged shuttle with the large and complex station. The relationship between the Cold War rivals of the space race had now become a collaboration. But there was little that NASA could do to fix their new friends' economic problems. Russia's space agency fell on hard times as the country stumbled out of Communism and tried to adapt to a new market economy. Mir could not survive on pride alone.
On June 25, 1997, Mir crewmen Aleksandr Lazutkin, Vasily Tsibliev and visiting NASA astronaut Michael Foale experienced a terrifying emergency. Tsibliev was using a semi-manual remote control system to guide an approaching Progress supply ferry towards a docking collar. It came in much too fast, failed to brake, missed the collar entirely and slammed into the side of a laboratory module (Spektr). The impact destroyed one of the solar arrays and pushed Mir out of its proper alignment with the Sun. Suddenly all the life-support systems' pumps and fans began to fail. Fresh air was no longer circulating properly within the modules and the three men were in danger of suffocating from their exhalations of carbon dioxide. The air pressure was dropping and there was an ominous hissing sound coming from the Spektr module.

Clearly the Progress ferry had torn a hole in the side of the module, but a hurried inspection failed to reveal exactly where the damage had occurred. A gash barely an inch long, hidden behind some wall-mounted equipment, was endangering the entire station. The three men scrambled out of the module and sealed the internal hatchway behind them. They had to cut through temporary electrical cables snaking through the hatch before they could shut it properly. If the puncture in the Spektr module had been any larger, they might not have had time to save themselves.

The problems piled up. Mir began to spin out of control, turning a complete cartwheel once every six minutes. Tsibliev and Lazutkin clambered into the Soyuz crew ferry and fired its thrusters, trying to

Cosmonaut Valri Korzun in the central node of Mir, where four modules meet. Space may be infinite, but living aboard a spacecraft is essentially like being sealed in a submarine.

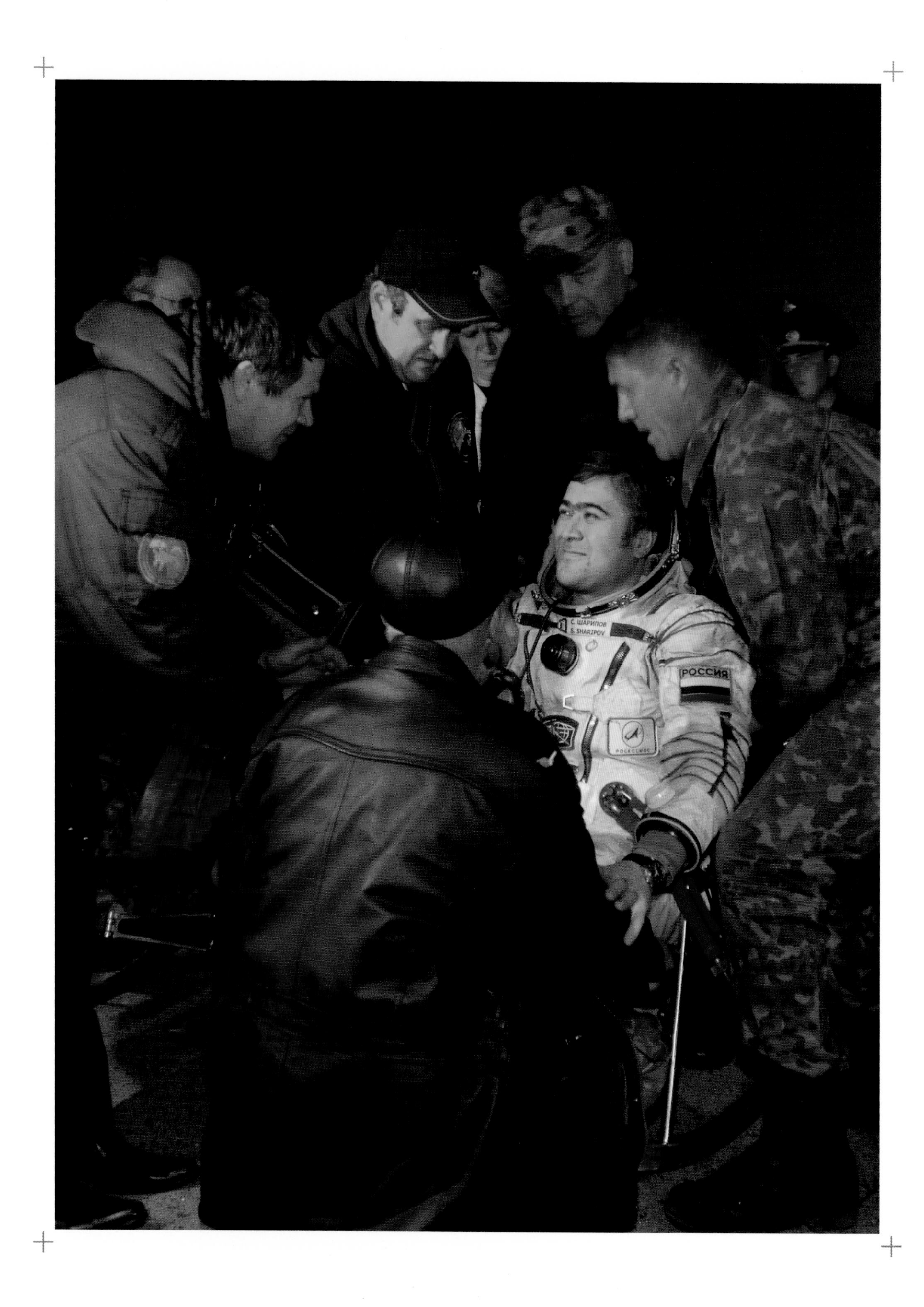
С. ШАРИПОВ
S. SHARIPOV
РОССИЯ

Aleksander Lazutkin and Vasily Tsibliev hug each other, relieved to be home safe after their troubled Mir mission. Opposite, Cosmonaut Salizhan Sharipov is welcomed back to earth in April 2005, on completion of his International Space Station tour of duty.

counteract the spin, while Foale, watching the earth's horizon from a small window inside Mir, made an educated guess about the spin rate and called out advice to his Russian companions through the open hatch of the docked Soyuz at the other end of the station. It was a confusing and terrifying experience, caused not by any failure of the crew but by a series of oversights and misunderstandings on the ground during an earlier redesign of the Progress ferry's guidance system. Foale endeared himself to his Russian colleagues, sharing the workload, helping to save the mission and loyally taking their side in subsequent enquiries. Lazutkin recalled Tsibliev's terrible guilt. "He was starting to worry about what the ground would do to him when he returned, because it was he who commanded the ship. I felt that he wasn't guilty of anything, but I couldn't find the facts to prove it. Michael felt the same."

When Tsibliev returned to earth in August, he was determined that neither he nor his crewmates should be blamed. He lost no time venting his fury to eager journalists. "It's been a long tradition here in Russia to look for scapegoats," he said. "The cause [of the crash] lies with problems on earth. It's connected with the economy, with our affairs in general. Even the items of equipment needed to live aboard the station, and that we asked to be sent—and we're not talking about coffee, tea and milk—they just don't exist. The factories don't work, or have insufficient supplies, or they ask for crazy prices." NASA was not entirely blameless either. Both agencies still had much to learn about properly supporting their people in orbit.

Tsibliev's reputation was soon restored and by 2000 he had been put in charge of cosmonaut training. But Mir's image did not recover. The headline-grabbing drama of the collision solidified an unfair public perception that the Russian space effort was dangerously antiquated. Several facts were overlooked by the world's media as they rushed to judgment. Mir survived an incredible 15 years in orbit. It pioneered the recycling of astronauts' biological waste and the robotic docking of new modules and uncrewed supply ferries (only one docking out of dozens went seriously amiss). It supported more than a hundred personnel representing fifteen different nationalities and paved the way for international cooperation, without which today's International Space Station could not have succeeded. Russian rockets, Soyuz capsules and Progress supply ships have provided essential backup for the ISS.

"The impact sends a deep shudder through the station.
It feels like a sharp, sudden tremor, a small earthquake ...
Diving headfirst into the module, he immediately hears an angry
hissing noise from somewhere below and to his left. It is, he
knows, the sound of air escaping into space. His heart sinks.
At this moment, Sasha Lazutkin is sure they are all about to die."

Bryan Burrough, *Dragonfly: NASA and the Crisis Aboard Mir, 1999*

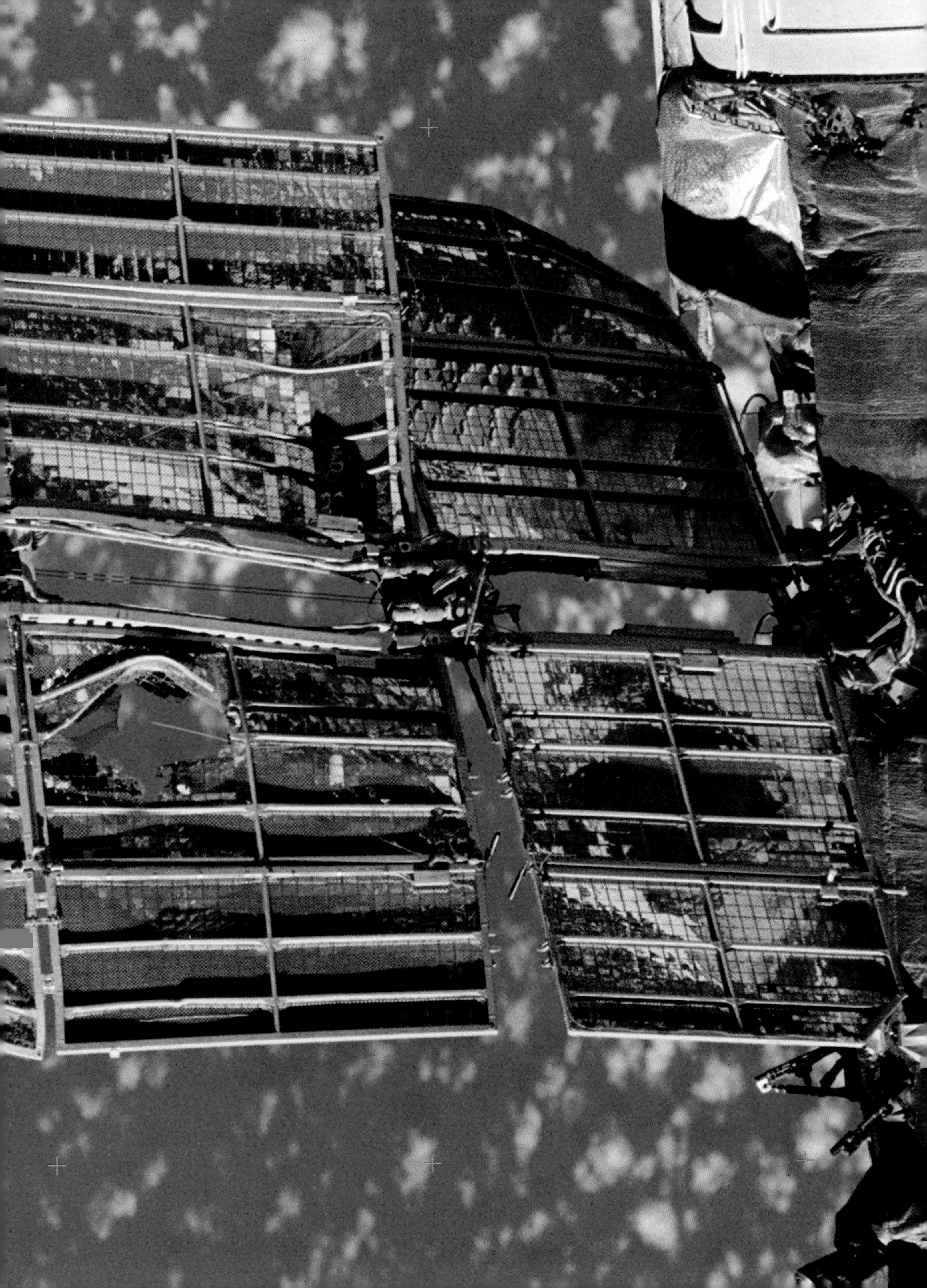

"Through the port hole I could see the heat shield burning away, getting thinner and thinner. I knew everything would be okay, but the shield was only about twenty centimeters from my head, and it was scary to see. On another flight, the parachute did not come out at the moment when it should have done, and I wondered what to do. And then it opened."
Cosmonaut Georgi Grechko, 1993

The Russian system of landing capsules on the ground rather than at sea has remained much the same for more than thirty years. In the last few moments before touchdown, retro rockets at the base of the Soyuz re-entry module automatically fire to soften the impact.

Space shuttle Discovery tumbles around so that space station crew members can check its heat shield tiles for damage.

Endeavour
2
1

Shuttle Discovery's underbelly of main reentry heat shield tiles, opposite, and above, the less critical white tiles on its upper fuselage.

This back has been necessary during prolonged periods when the shuttle has been grounded—for NASA's flagship spacecraft has also won an undesired reputation as a troublesome and antique machine. While its successes are undeniable, its failures have been far more terrible than any of Mir's little dramas.

In the late 1990s, New York photographer Adam Bartos traveled to a number of Russian space facilities and was granted adequate time to create painstaking photographs.

"I was very impressed by what I saw. The lack of resources makes the accomplishments that much more striking. I did get a sense of the pride people take in their work, and their interest in posterity is evident. Every company has a museum and sometimes schoolchildren visit these. Many times in the U.S., when I've told people what my photographs are about, they immediately say something disparaging, expressing surprise that the Russian program still exists, or making a quip about tin cans and so on, which is sad, and misinformed."

Neither can it be claimed that American technology has been a paragon of perfection. The story of the space shuttle has been troubled from its earliest days.

"People forget that the space shuttle is always flying at the edge of the envelope."
The Boston Globe, February 3, 2003

Press photographers capture a shuttle launch. A smooth lift-off wins relatively little space on TV and newspaper reports. It is assumed that space flight should be routine unless something goes wrong. In fact it never is. Every mission is essentially a hazardous experiment in this relatively new human adventure.

Flawed Flagship

> When commander John Young and pilot Robert Crippen strapped themselves into the space shuttle Columbia for its inaugural flight on April 10, 1981, it had been six years since America last sent astronauts into orbit. The shuttle's development had taken far longer—and cost far more—than NASA had originally expected. Inside the cabin, five computers cross-checked each other's results. For safety, at least four of them had to match results before Columbia could fly, but twenty minutes before the scheduled lift-off, they were not in accord. Young and Crippen had to climb out and go home. A million people camped out on nearby beaches were told they would have to pitch their tents for another couple of days to see the launch.

At last, on April 12, Columbia lifted off the pad and ascended flawlessly. NASA was back in business and public reaction was ecstatic. Once in orbit, Young and Crippen experienced the benefits of Columbia's large cabin, which was much roomier than the old Apollo capsules' had been. There was even a second deck beneath the cockpit with an airlock, sleeping berths and a private washroom cubicle. Behind the cabin was a payload bay about 65 feet long and 15 feet wide. One day soon it would carry pressurized space laboratories, satellites or planetary probes, but right now it was empty.

For this first flight, Young and Crippen were strapped into ejection seats. Once the shuttle had proven reliable, these would be replaced with seven crew couches (four on the main deck and three down below). But was the ship safe yet? When the two men looked out of the rear-facing cargo bay windows, they were in for a shock. Sixteen heat shield tiles had fallen off during launch and as it transpired later, at least a hundred more were slightly damaged. According to Crippen, "Our own engineers said that a few of the tiles on our underside had probably fallen off. If enough had, they said, then the plasma heat from our reentry would burn right through us."

NASA asked the US Air Force to point one of its secret spy satellites at the shuttle while it presented its vulnerable underbelly for inspection. The photos were never released, although it was made clear to NASA that none of the most crucial tiles was missing.

After two days in space, Columbia headed back to earth. Crippen described the moment: "We hit the atmosphere at around 24 times the speed of sound. After a while, we could see a pink glow outside our windows, but other than that, there absolutely was no sensation of heat."

Columbia then used its large wings to glide, unpowered, to a perfect landing at Edwards Air Force Base in the Californian desert: the legendary testing ground for new aerospace vehicles. Unlike with a conventional plane, there was no way to turn around during the final approach. Columbia had to touch down correctly on the first try, guided by its computers and the occasional nudge from a human hand on the controls. Crippen was thrilled. "The shuttle was a joy to

"For a successful technology, reality must take precedence over public relations, for Nature cannot be fooled."
Physicist Richard Feynman, 1986

fly. It did exactly what you told it to, even in unstable regions of air." It looked like a typically smooth NASA adventure and unlike many space missions today, Columbia's debut made headlines around the world. Behind all the smiles and the flag-waving, however, the story had not been such a happy one.

When Thomas O. Paine became NASA's third chief in 1969 he discovered that the U.S. president at that time, Richard Nixon, had little interest in space. Nixon even talked about canceling manned flights in the coming decade. Paine knew that Apollo-style rockets were too expensive to fly because they were thrown away after every mission. He asked Nixon for $12 billion to fund a space plane, a "shuttle" that would piggy-back on a separate winged booster. Both components would be fully reusable. They would be expensive to build, but cheap to operate for many years after. The shuttle would carry components for a space station, and then ferry crews back and forth. After all, what was the point of a shuttle with no destination?

The shuttle's future is limited. NASA is turning its back on winged space ships.

Nixon was unimpressed. The space station was put on indefinite hold and NASA revised its shuttle plans. A cheaper version would now pay its way by carrying commercial satellites into orbit, turning the once-proud NASA into a trucking company. Still Nixon demanded an even lower budget. Humiliated, Paine resigned. "Obviously I can't deliver the kind of relationship with the President that NASA deserves," he commented sadly.

Republican budget adviser (and Nixon's future Secretary of Defense) Caspar Weinberger intervened. On August 12, 1971, he sent Nixon a short memo that probably rescued the shuttle from cancellation. "We don't want the world to think our best years in space are behind us. NASA's proposals have some merit.

Space shuttles are complicated and expensive to operate. They require thousands of staff to service and make ready for launch. When the shuttle retires from flight in the year 2010, NASA faces a stark choice. A complete reinvention of America's rocket technology might seem the best course of action, but thousands of jobs are

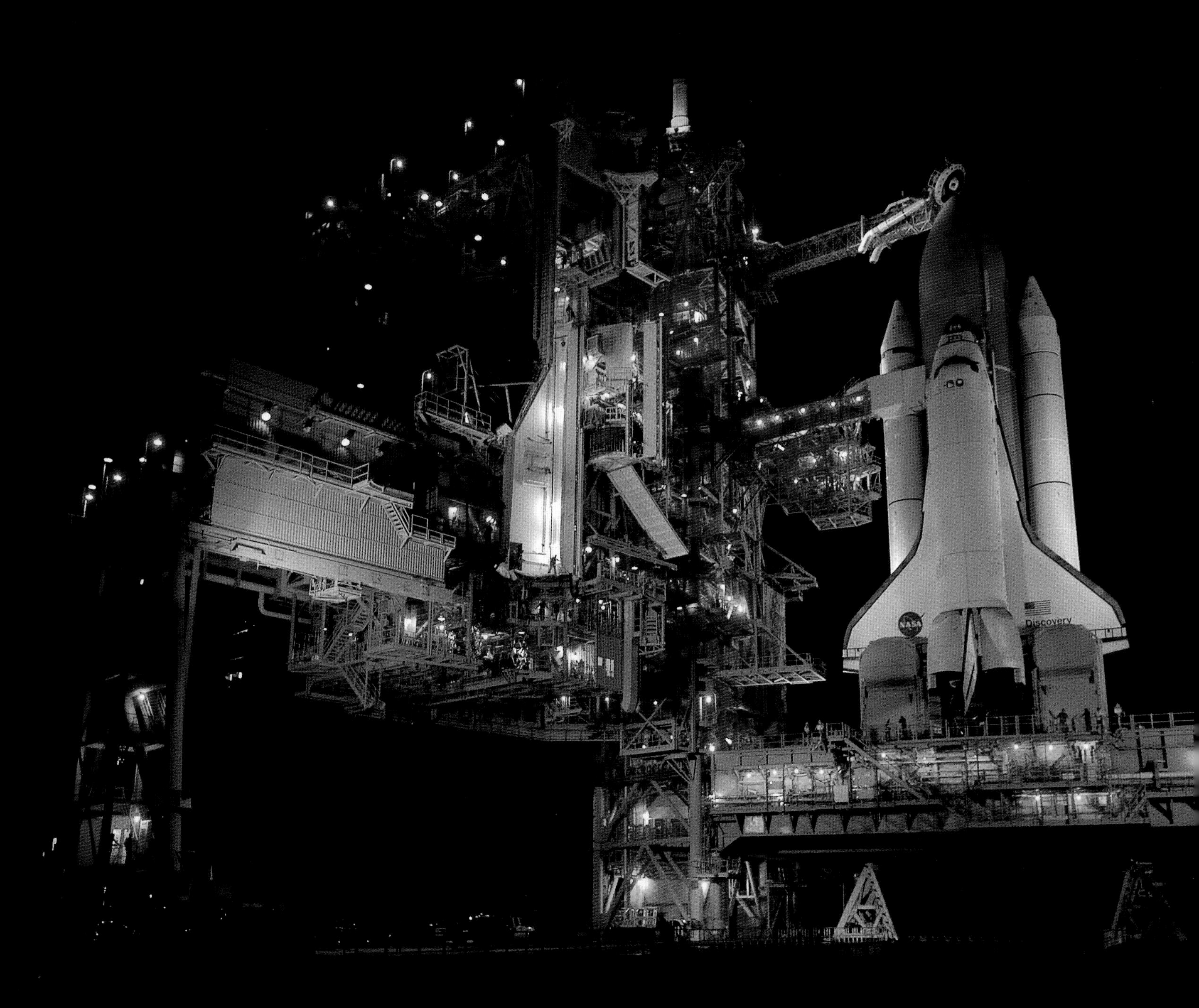

"We need to make things as simple, as clean, and as automated as you can in the vehicle design and in the ground processing, so we can get about our business in a more efficient manner than before."

NASA Chief Administrator Michael Griffin, November 2005

A successful launch thrills the crowd. Jeb Bush, governor of Florida (in the blue shirt), watches with President Bush's wife Laura Bush on his right, above. Opposite, failures cause heartbreak and loss. Space flight is about people, not just machines.

"Obviously a major malfunction..."
Mission Control, January 28, 1986

"There once was an organization that was regarded as being perfect, that suddenly doesn't do the simplest thing. My point of view was that you had to do all the simple things as well as the complicated things if you are going to succeed."

Former NASA chief James Webb, 1986

Challenger's lost crew: in the back row, from left to right, are Mission Specialist Ellison S. Onizuka, Teacher in Space participant Sharon Christa McAuliffe, Payload Specialist Greg Jarvis and Mission Specialist Judy Resnik. In the front row, from left to right, are Pilot Mike Smith, Commander Dick Scobee and Mission Specialist Ron McNair.

"I want to say something to the schoolchildren of America who were watching the live coverage of the shuttle's take-off. I know it's hard to understand, but sometimes painful things like this happen. It's all part of the process of exploration and discovery. The future doesn't belong to the fainthearted. It belongs to the brave."

President Ronald Reagan, January 28, 1986

We keep slashing their budget because we can, and not necessarily because they are doing a bad job." Nixon's response was to scribble a quick handwritten note in the margins of the memo. "I agree with Cap."

Even so, NASA was only awarded $5 billion for the shuttle: less than half of the projected costs. The reusable winged carrier stage had to go and the design now included partially reusable solid rocket boosters that were little more than over-sized fireworks, along with a liquid fuel tank that would be discarded after each use. From the start, the shuttle was half the ship it should have been.

New engine technology had to be developed on a shoestring, stretching ingenuity to the limit. Very soon NASA found itself spending more than its $5 billion limit. Dan Dumbacher was on the team developing the shuttle's main engines. "It was not an easy problem to make engines that could last many flights instead of just surviving a few minutes and then getting thrown away forever into the Atlantic. First, you have super-cold liquid hydrogen and oxygen at minus 400 degrees Fahrenheit flooding in. Then, fractions of a second later, you're combusting these fuels at plus 5000 degrees F. You're lucky if you can keep the engine's outer structure at a modest 1000 degrees F to stop it melting. And you have to force-feed fuel at enormous pressure. The turbine fuel pumps have blades that spin at 35,000 rpm. That's twice the speed of a jet engine."

The heat shield tiles were another nightmare. Some 34,000 ceramic panels had to be glued onto 70 per-cent of the shuttle's outer surface. Because of the ship's aerodynamic curves, no two tiles were shaped alike. As these and other problems piled up and the years went by, the limited funds ran out and NASA had to ask a new president, Jimmy Carter (again no fan of space), for more money. It was a case of spend, spend, spend, and still nothing to fly.

But at last, in 1981, Columbia's launch seemed to banish these shadows. Newly elected president Ronald Reagan had just trounced Carter at the polls. He congratulated NASA and happily exploited the rhetoric of space. Five years went by, sister ships Challenger and Atlantis went into service and all three shuttles achieved dazzling successes in orbit during what appeared to be an increasingly routine schedule. Then, on its 25th mission, Challenger exploded on January 28, 1986. All those compromises from the Nixon era were exposed in a series of design flaws.

One of Challenger's solid rocket boosters sprang a leak. Freezing weather had affected the rubber rings that sealed the various sections. A jet of flame escaping from the faulty seal burnt into the liquid fuel tank and 73 seconds into the flight the entire ship blew up. Six astronauts and one civilian teacher, Christa McAuliffe, were killed. The shuttle fleet was grounded for an intense and emotionally grueling investigation.

The renowned physicist Richard Feynman was invited to join the board of inquiry, the Rogers Commission. He immediately cleared his own eccentric path through the verbiage. For instance, with the TV cameras

NASA
GODDARD
SPACE
LOCKHEED MARTIN

"Despite New Data, Mysteries of Creation Persist."
New York Times story headline, May 12, 1992

A *Time* magazine cover highlights the nationwide sense of doubt in the wake of the Challenger tragedy. After a troubled start to its career in 1990, the Hubble Space Telescope, opposite, eventually helped rescue NASA's beleaguered reputation.

running, he dipped a piece of rubber into a glass of iced water and showed how it hardened when cold.

"Do you suppose this might have some relevance to our problem?" he asked, knowing very well that it did. He had created a vivid demonstration of the notorious "O" ring flaw that had destroyed the shuttle during its wintry launch. But Feynman wanted to probe further.

"If NASA was slipshod about the leaking rubber seals on the solid rockets, what would we find if we looked at the liquid-fuelled engines and all the other parts that make up a shuttle?"

Feynman was told that it was beyond the inquiry's brief to look at the main engines because no problems had been reported. So he made unauthorized trips to NASA facilities where he could speak to ground-floor engineers in private. He wrote later, "I had the definite impression that senior managers were allowing errors that the shuttle wasn't designed to cope with, while junior engineers were screaming for help and being ignored." He had identified a serious human problem that haunts NASA to this day.

Feynman also looked at the complex relationships between the space agency's many departments and their private industrial suppliers. "NASA's propulsion office in Huntsville designs the engines, Rocketdyne in California builds them, Lockheed writes the instructions and the launch center in Florida installs them. It may be a genius system of organization, but it seems a complete fuzzdazzle to me."

In the last days of the inquiry, Feynman made a plea for greater realism. "For a successful technology, reality must take precedence over public relations, because Nature cannot be fooled."

For this generation, the Challenger catastrophe defined NASA's public image as a complicated, flawed agency, yet it can always be relied upon to come back from disaster and flash its old-time "can-do" smile. Once the shuttle had returned to flight, a particular mission caught the public imagination. It was widely hailed as among the most worthwhile projects ever undertaken by astronauts.

Regaining Trust

Astronaut Story Musgrave, anchored on space shuttle Endeavor's robotic arm, during the first Hubble servicing mission in December 1993.

> After its launch in April 1990, the Hubble Space Telescope had begun its operational career in disarray. Its main mirror was manufactured to the wrong specification, so the telescope that was supposed to give us our sharpest-ever view of the farthest reaches of space was essentially blind. Just as embarrassing, the huge solar panels that powered the onboard equipment flexed and buckled every time Hubble's orbit took it into, or out of, direct exposure to the Sun. The $2-billion machine, a decade in development, was apparently a farce. Fortunately, the telescope had been designed right from the start for in-orbit servicing by shuttle astronauts. Optical engineers realized that the faulty mirror could be corrected at the "eyepiece" end of the assembly, using a clever system of lenses to bend the fuzzy images back into focus. Meanwhile, the British manufactur of the solar panels worked out why they were behaving so poorly and built a new pair that would remain stable during temperature fluctuations.

On December 2, 1993, shuttle Endeavor lifted off carrying a rescue crew. Working as two separate duty teams, astronauts Story Musgrave, Jeffrey Hoffman, Thomas Akers and Kathryn Thornton completed the repairs in five back-to-back spacewalks totaling 35 hours. Hubble was pulled into Endeavor's cargo bay using the Canadian-built Remote Manipulator robot arm, a versatile piece of equipment that has featured in many shuttle flights. Musgrave and Hoffman secured Hubble and began replacing equipment. Then Thornton performed a dangerous but necessary maneuver, using a pair of shears to snip off the old solar panels, while Endeavor produced a brief burst of thruster fire to send the panels safely down into the atmosphere.

Next, Thornton and Akers loaded a self-contained package of corrective optics into a side hatch on Hubble. During subsequent spacewalks, the crew added more instruments and new gyroscope control systems before gently easing Hubble free of the Endeavor and back into an independent orbit. Hubble's instruments may have been flawed, but the main spacecraft structure turned out to have been well designed. The access doors opened properly, and the new equipment slid into place according to plan. The spacewalks were spectacular, and NASA's regained confidence seemed infectious. Some of the agency's congressional critics began to revise their views, perhaps unwilling to appear ungracious in the light of an obvious success. The complex spacewalks also demonstrated that building a space station need not be as foolhardy a scheme as some commentators had imagined.

The shuttle has certainly achieved great things and is a true marvel of engineering. Yet in the early days when NASA and the world's press celebrated it as "the most complex machine in history", few anticipated that this complexity would prove so dangerous. The shuttle has been a science fiction dream come true, an awesome winged spaceship flying time and again into orbit and coming home to land like a plane. But sometimes, like so many science fiction dreams come true, it has delivered nightmares as well.

JPL

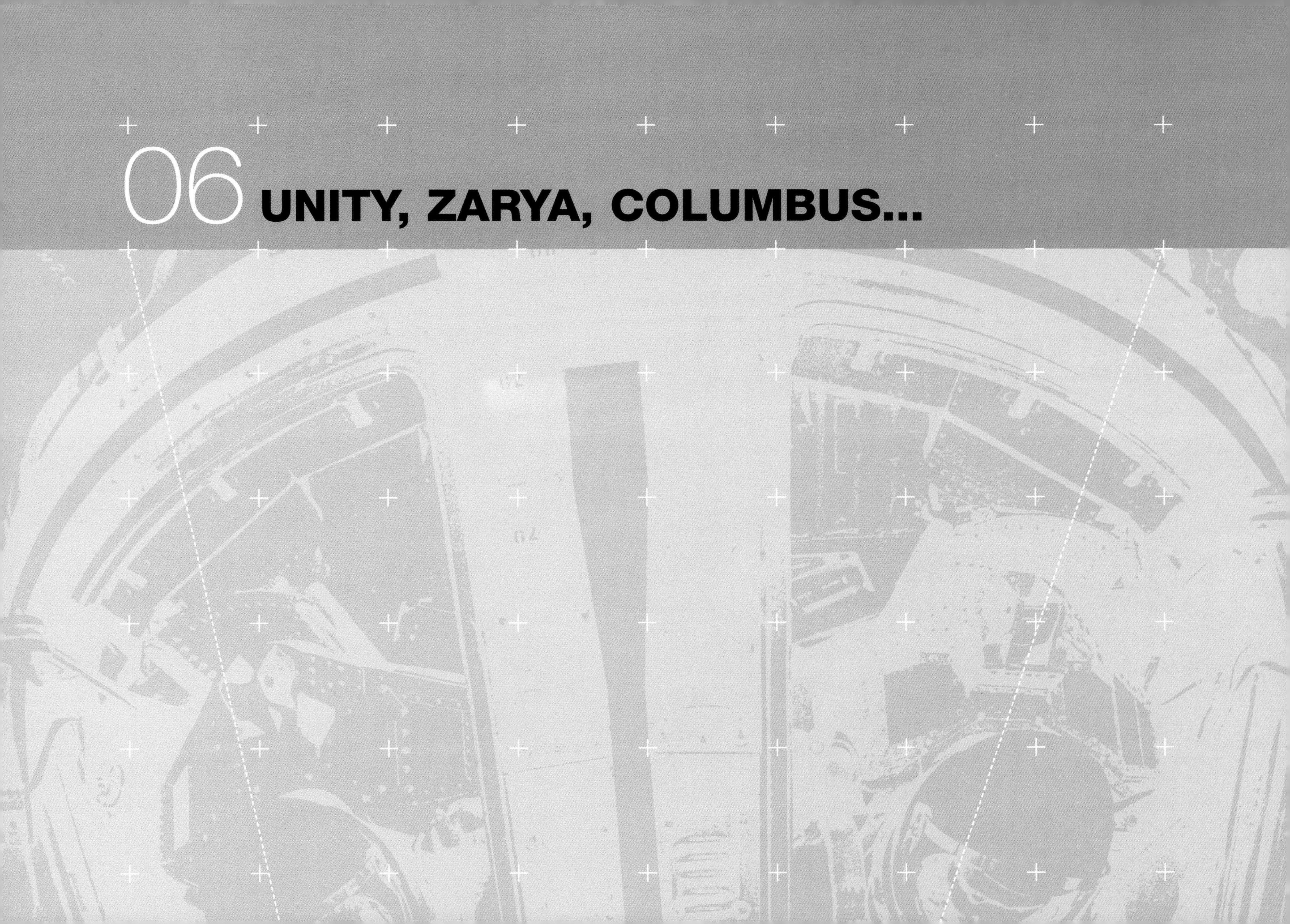

06 UNITY, ZARYA, COLUMBUS...

"There was some debate about Russian dependability, suggesting that their economy and engineering could not be counted on. It's turned out that they have been very reliable, really quite strong compared to us."

Charles Vest, President, Massachusetts Institute of Technology (MIT), 2003

UNITY, ZARYA, COLUMBUS . . .

> The space race may have ended with NASA's landing on the moon, but that does not quite explain how America and Russia became such solid allies in orbit. Even twenty years ago, the idea of an American astronaut relying on a Russian craft for a ride into space would have seemed unthinkable. Today, the astronaut program depends on the close cooperation of the two nations.

On June 6, 1992, NASA chief Dan Goldin and his Russian opposite number Yuri Koptev met for the first time in Goldin's Washington apartment. Their meeting was not exactly secret, but neither was it reported in the press. Such caution was understandable. At this early stage, neither man knew what he would find in the other. Despite barriers of nationality, history and language, Koptev and Goldin quickly saw how similar their problems were. They were both former Cold Warriors who believed in exploring space for peaceful ends. They both had been stripped of much power by their political masters and forced into painful spending cuts. Space historian John Logsdon describes this meeting as "like the two of them falling into 'administrator's love'. They were almost mirror images of each other."

Goldin and Koptev's meeting was not entirely without diplomatic precedent. Kennedy and Khrushchev flirted with the idea of stepping back from competition in space, and since the collapse of the Soviet regime, Russian and American leaders have often discussed joint space activities at their regular Summit meetings. Richard Nixon welcomed the docking of NASA's last Apollo craft with a Russian Soyuz in 1975 as a signal that the space race was drawing to a close. Both

NASDA

Previous page, a computer graphic shows the International Space Station as it might have looked today, if all had gone according to plan over the last decade. The real station is not so large. Opposite, a wrecked Soyuz rocket booster segment lies abandoned in a Kazakh field. Unburned fuel and toxic chemicals spill into the soil, angering the nearby farm communities.

Jimmy Carter and Ronald Reagan considered inviting the Soviet Union to participate in a simulated rescue mission between a shuttle and Salyut 7, the last of Mir's predecessors. At that time, the Soviets did not feel that their crews would ever need rescuing by NASA, and little more progress was made. In fact Leonid Brezhnev, one of the last old-style Soviet leaders, had poured vast amounts of money into Buran, a largely pointless look-alike competitor for NASA's shuttle that flew only once before being abandoned.

A decade later the situation was very different. Post-Soviet Russia was now thoroughly pre-occupied with getting rescued, not so much through diplomatic dockings but with dollars. Within days of Goldin and Koptev's private meeting, Presidents George Bush senior and Boris Yeltsin agreed to make plans for sending an American shuttle up to Russia's Mir space station and also to fly Russian cosmonauts on a shuttle. This was a real diplomatic advance, yielding the first hope for joint spaceflights since the Apollo-Soyuz docking of 1975.

At its height, the Russian space program had employed 600,000 people, many of them wearing military uniforms. The Soyuz vehicles, the Salyut stations and Mir had contributed to an almost non-stop presence in space and by 1991 Soviet cosmonauts had clocked up twenty-three years' worth of flight time between them, while their American rivals had accumulated just eight years.

"We are 225 miles closer to the stars. We are humanity's only outpost in space at this time. That is something very special."
Space Station Expedition 10, Commander Leroy Chiao, 2004

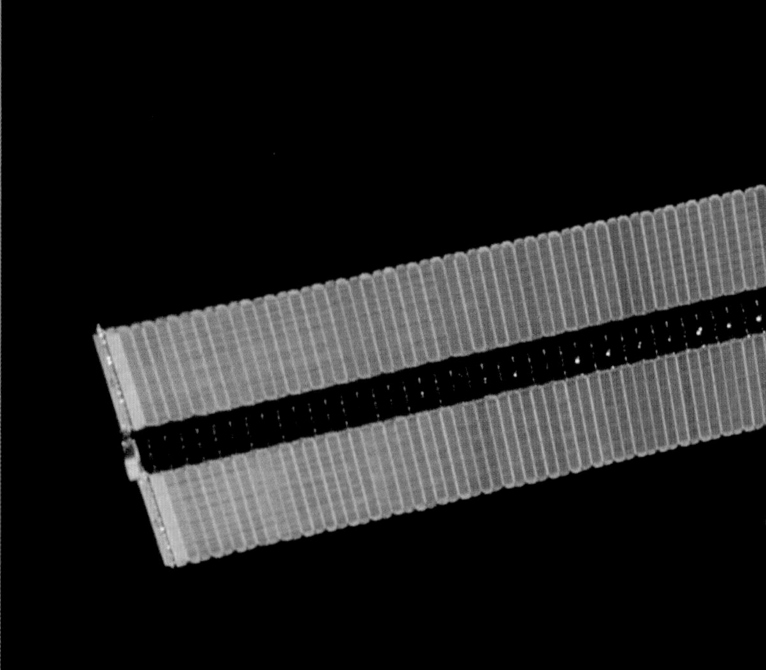

Previous page, the International Space Station's docking hatch in the last few seconds before Shuttle Atlantis locks into place.

Opposite, the International Space Station seen from a slightly lower orbital altitude.

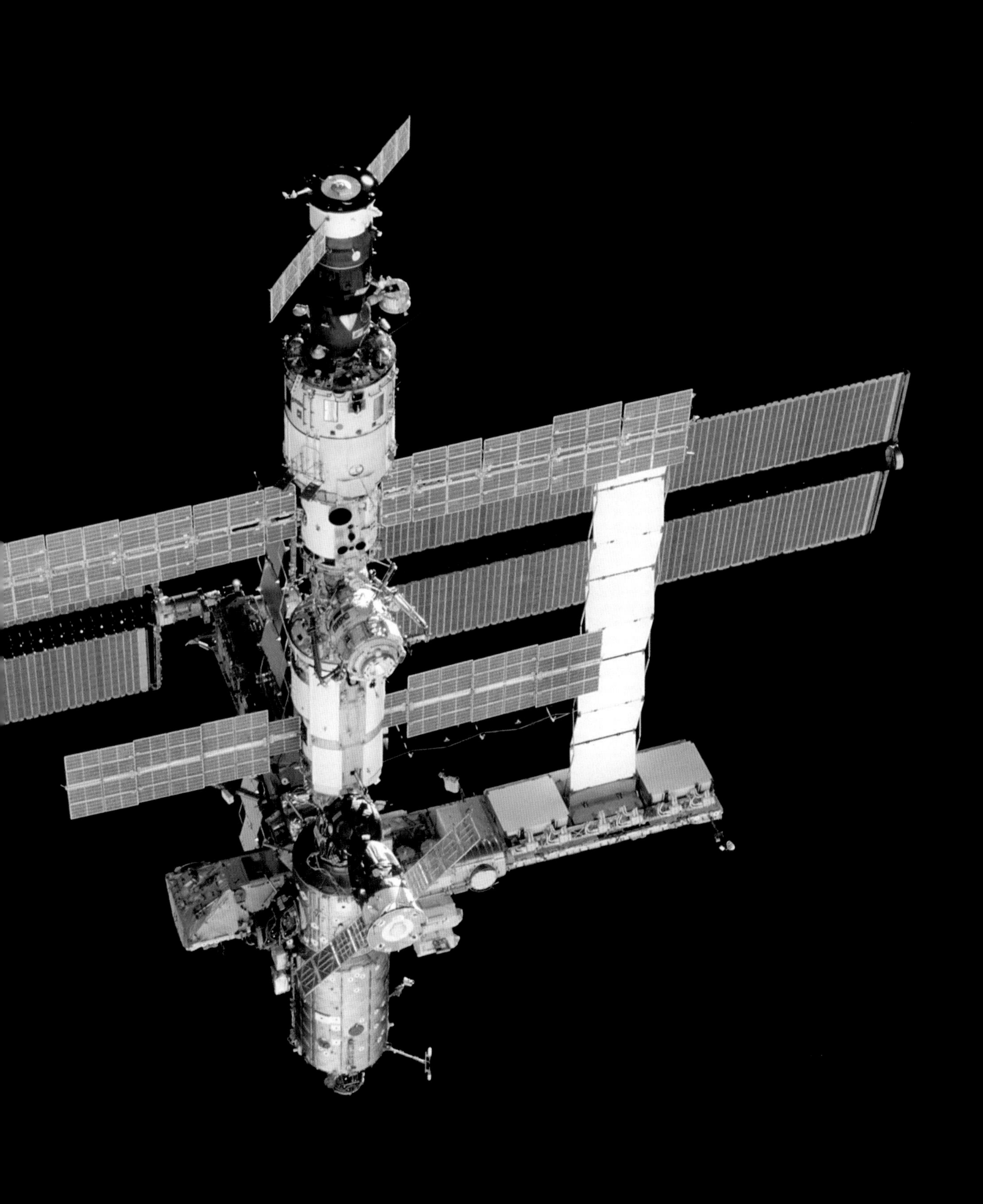

Despite escalating economic problems, Soviet rockets, carrying unmanned probes, earth satellites and Progress ferries, were launched at the blistering pace of one every week.

Decay set in rapidly, however. In May 1991, the Mir station had been operational for five years and was starting to creak. Flight engineer Sergei Krikalev was sent up on a six-month tour of duty. He conducted a number of spacewalks and worked through a busy schedule of repairs. By October, he was looking forward to the docking of another Soyuz bearing a relief engineer so he could go home to his wife and baby daughter. But some of the messages from ground control were making Krikalev uneasy. They explained that while a special guest astronaut was scheduled to visit Mir in November, there wasn't yet enough money for that flight. So they would bring up the guest on the October relief mission, along with an Austrian astronaut, Franz Viehbock, whose government had contributed $7 million towards the flight. There would be no replacement engineer for now. Krikalev would be staying aboard Mir for another five months.

Krikalev was used to guest astronauts. On his flight up to Mir he had sat next to British astronaut Helen Sharman, who had stayed on the station for a few days, subsidized by Russia as an experimental act of goodwill after the London-based private consortium that was supposed to pay for her flight failed to raise the promised money. More than a dozen foreigners

Astronaut James Voss at work in the Space Station's Destiny science module, while shuttle commander Scott Horowitz floats through the hatchway during August 2001.

had visited Mir so far, including a Japanese journalist and serious-minded European astronauts who conducted genuine science programs. Ticket prices for these visitors ranged from $8 million to $12 million, depending on their length of stay. Gradually it became almost impossible to schedule any Soyuz launches unless a foreigner was paying for the flight.

But this latest guest was not quite as foreign as Krikalev might have expected, and neither would he be paying a fare for the ride. The space authorities in Russia had lost control over their remote launch site at Baikonur in Kazakhstan because that once-tame republic in the south of the Soviet empire had shrugged off Moscow's dominion and declared itself an independent country. Kazakh authorities were now demanding that Russia make the goodwill gesture of sending a Kazakh citizen, Toktar Aubakirov,to Mir at the first opportunity, or else Russian space workers would be denied further use of the Baikonur complex.

ESA astronaut Pedro Duque, in a mock-up of Russia's Zvezda Service Module, training for his mission in 2003.

And there was more unsettling news. The mission controllers weren't too sure how to tell Krikalev about the tanks that had rumbled through the streets of Moscow that August during a coup attempt that ultimately unseated the liberal reformer Mikhail Gorbachev but failed to reinstate the old regime as had been intended. In fact the coup had the opposite effect, destroying the last vestiges of communist rule and shattering the Soviet empire. Krikalev's mission badge proclaimed him a cosmonaut of the Union of Soviet Socialist Republics. As from now, that place or state of mind no longer had any sensible meaning. Krikalev was profoundly adrift.

"The biggest developments of the immediate future will take place, not on the moon or Mars, but on earth, and it is inner space, not outer, that needs to be explored. Even in space, the most alien creatures we'll confront are ourselves."
Author J.G. Ballard, 2004

Neither did mission control highlight the fact that two of the coup's leaders, Oleg Baklanov and Vitaly Doguzhiyev, had been senior figures in the rocket program. With this in mind, Russia's new leader, Boris Yeltsin, had become ill-disposed to the space program. In February 1992 he abolished Krikalev's employer, the once-great Ministry of General Machine Building, replacing it with smaller and less powerful entities. Rocket activities were now the responsibility of a new specialized authority, the Russian Space Agency (RSA). Commercial activities, the hunt for foreign hard currency, now had precedence over the old military posturing. Immediately the RSA was plunged into the same kind of struggle that NASA had faced: to keep the army generals from hijacking their dwindling financial resources.

When Krikalev finally came home in March 1992, he had spent 313 days on a tiny Communist-built island in space while his vast homeland tore itself to pieces and changed forever. He found to his dismay that the back salary for his long and exhausting mission was a pitiful reward for his services, equating in near-worthless roubles to maybe a couple of hundred American dollars for ten months' work. He and his family could barely afford food. Krikalev's new employer, the RSA, was in no position to help, since its finances had collapsed before it could even get started. None of its employees could afford food. A colleague of Krikalev's summed up the situation to a Western journalist: "Once the government would throw more money at us than we could eat. Now we wonder about our next meal."

The jubilant Kazakh authorities had gained a partial stranglehold over a rapidly diminishing asset: the Baikonur launch center itself. The nearby town of Leninsk had grown up around the center over the past 40 years and now tens of thousands of its citizens, mainly skilled Russian and Ukrainian technical staff, were packing up their possessions and moving away like war-torn refugees, because Baikonur no longer had jobs for them. The remaining Kazakhs had always been treated as second-class citizens in their own land. Russia had never trained them in the engineering skills required to launch rockets, or even to keep Leninsk's power station going.

Resentments boiled over. Discarded first-stage boosters from four decades of Russian launch activities were scattered across vast areas of Kazakh farmland, leaking unburnt fuel and other toxic chemicals into the ground. Under Soviet rule, nobody had dared complain. Now the locals demanded that Russia clear up this rubbish and surrender half its ownership of Baikonur. Meanwhile, the newly instated Kazakh border customs hijacked Russian machinery on its way to the launch center and held much of it for ransom.

In February 1992, just as the newly founded Russian Space Agency (RSA) was trying to forge some kind of working relationship with the Kazakhs, Baikonur's army guardians rioted in the first of several rebellions over lack of food, poor salaries and the squalor of their living conditions. They burned their barracks to the ground, and three people died in the chaos.

Wearing a Russian Orlan spacesuit, NASA astronaut John Phillips works outside the space station in August 2005.

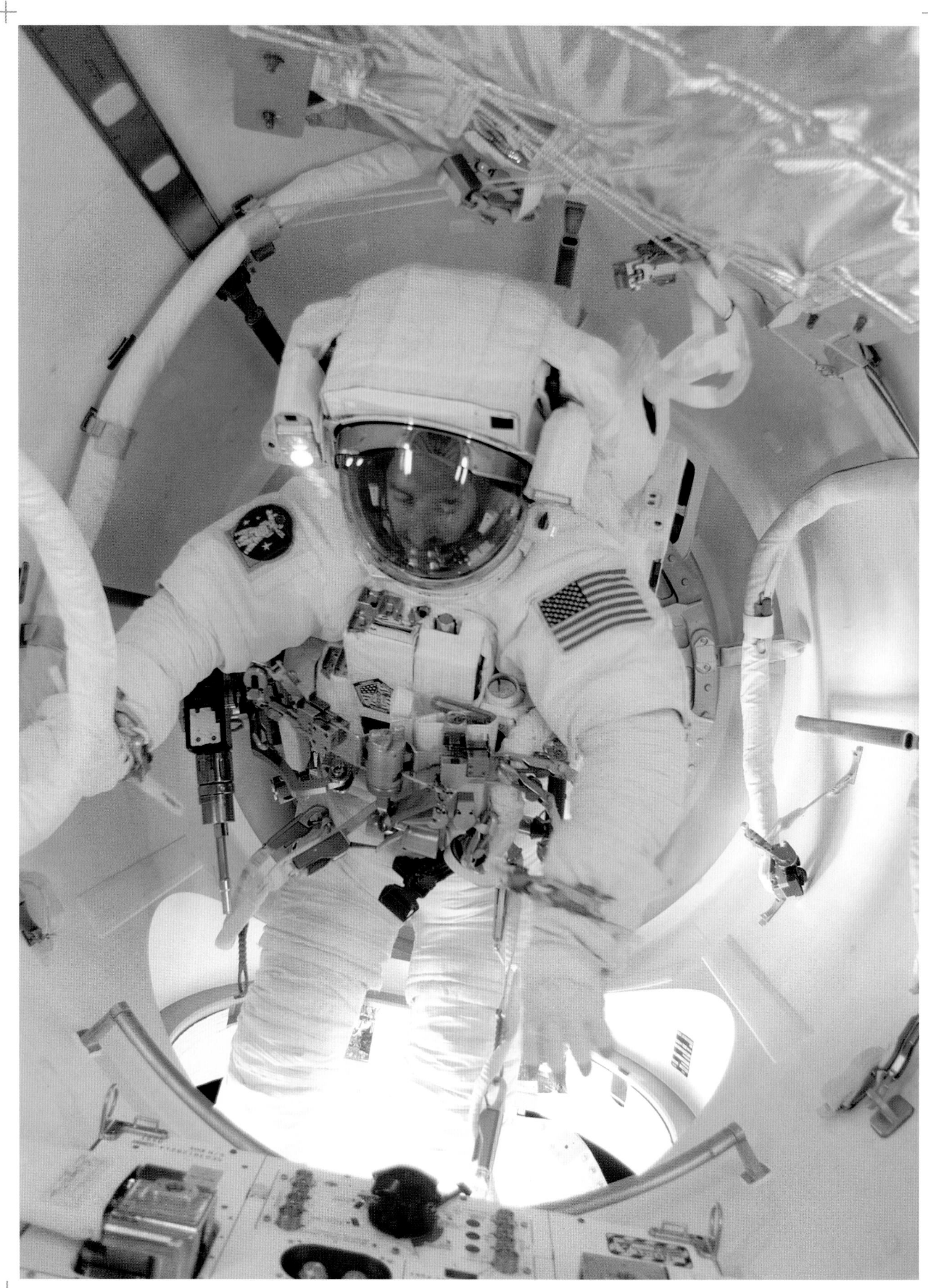

Page 218, Astronaut James F. Reilly uses the International Space Station's airlock for the very first time in July 2001. Page 219, only those who have flown into space can truly understand the exhilaration of weightlessness, and the wonder of seeing the earth from orbit.

The European Space Agency ESA is a major contributor to the space station project. This graphic, opposite, depicts the Columbus science module, currently under construction and awaiting launch.

A year later, the RSA's problems at Baikonur had become catastrophic. In May 1993, a rocket bearing a communications satellite fell into the Pacific like a damp squib because there hadn't been enough clean fuel available for Baikonur's gantry crews to fill the booster's tanks. A month later, 500 Kazakh conscripts stormed the center, smashing equipment, hijacking Russian workers' cars and causing massive damage.

Russia's new president Boris Yeltsin recognized that foreign cash investments in Russia's remaining space capabilities could only be threatened by this chaos. On March 28, 1994, he signed an agreement effectively renting Baikonur from Kazakhstan for the equivalent of $110 million a year, paid in the form of trade credits. A few weeks later, a fire at the center destroyed rocket components in an assembly shed. There was no water available to fight the flames.

It is hardly fair to compare subjugated nations like Kazakhstan with any Western equivalent, but imagine a parallel: suppose California, Texas, Florida and all the other American states suddenly declared independence, while a failed ultra-Republican coup forced tanks onto the streets of Washington D.C. Then expand the usual in-fighting between NASA and the military and throw in the conflicting interests of the many rival contractor companies to produce a mélange of bickering and confusion.

Imagine NASA broken up to neutralize its political influence and the various field centers now competing for control over what little space activity remains. Imagine persistent rioting and looting at the Kennedy launch center, while Floridian authorities impose heavy taxes on hardware arriving from other states to be assembled for flight. Imagine Florida then demanding that the Johnson control center in Texas pay a massive levy to use the launch center at all. Add to this the poison of near-endemic bankruptcy, causing thousands of job losses and plunging the value of space workers' salaries from $30,000 a year to maybe a couple of thousand dollars at best, and the effective running of a space agency becomes impossible.

How low might the morale of your space workers sink under such circumstances? A rocket ferry bound for a space station might, for instance, be looted on the launchpad for its food packages by ground crews desperate to feed themselves, or stripped of its contents to sell on the black market for hard currency. Imagine the ferry blasting into the sky half emptied of its supplies. This is exactly what happened at Baikonur in mid-1994 when a Progress capsule going up to Mir was pillaged just before launch.

Yuri Koptev had the unenviable task of trying to sort out this mess. He had been second in command at the old Ministry of Machine Building but was untainted by any association with the failed coup. Now he headed the new Russian Space Agency. This big, burly man with a gruff manner and a steely gaze was perhaps the most important unifying factor as the space program struggled to reassemble itself into a working whole.

NASA

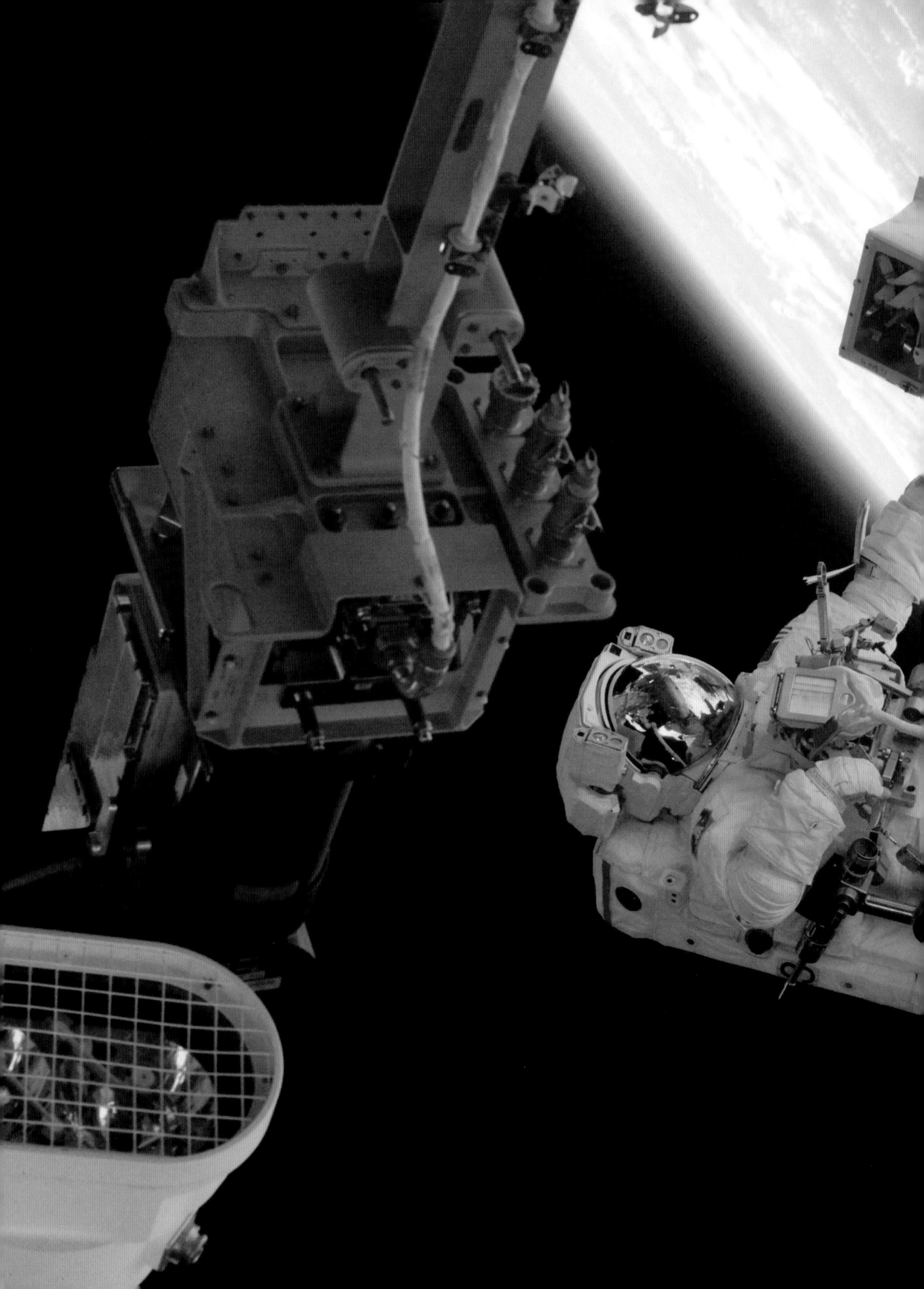

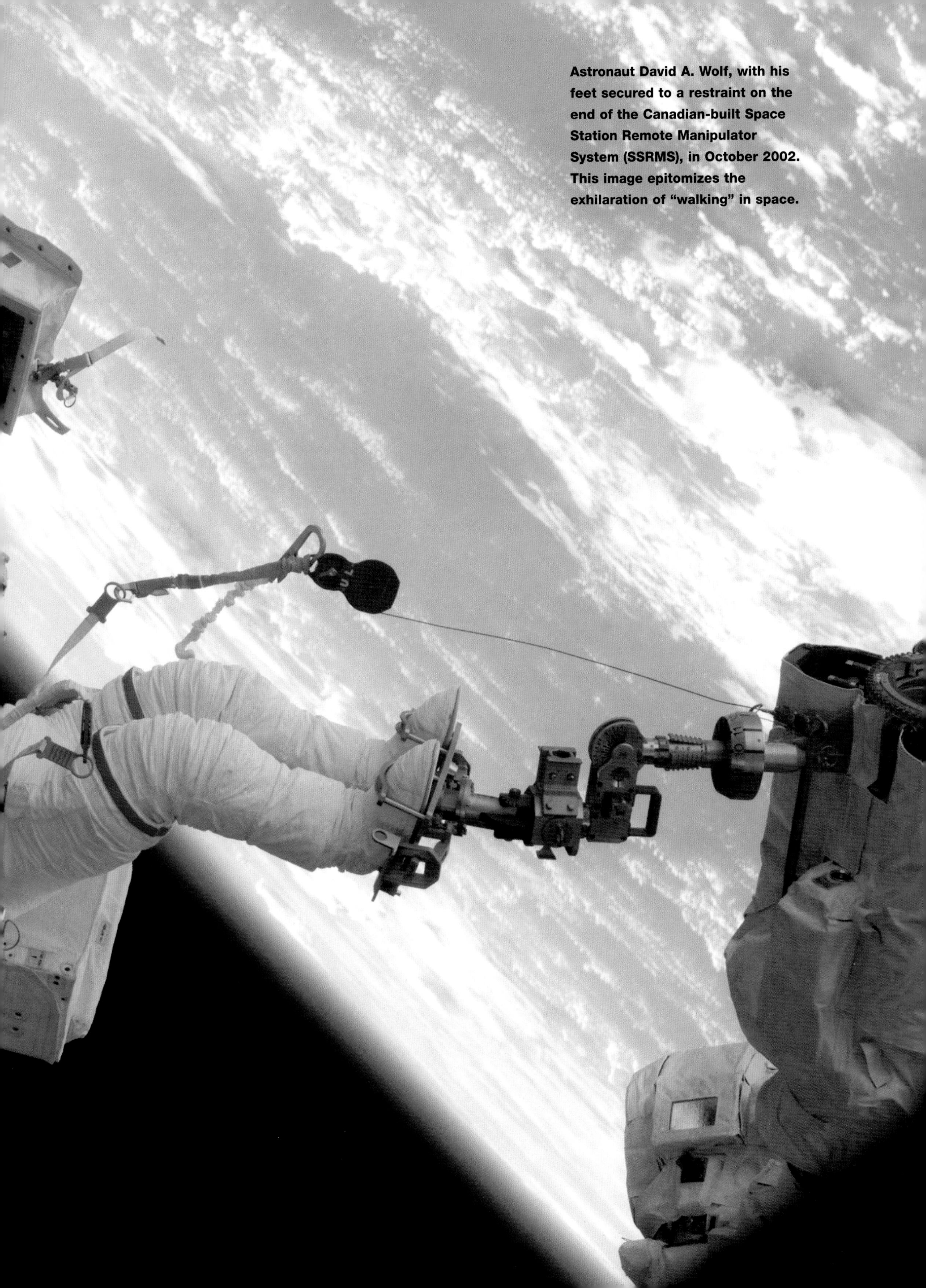

Astronaut David A. Wolf, with his feet secured to a restraint on the end of the Canadian-built Space Station Remote Manipulator System (SSRMS), in October 2002. This image epitomizes the exhilaration of "walking" in space.

"Building just one space station for everyone was insane. We should have built a dozen."
Science fiction author Larry Niven, 2000

Koptev encouraged American aerospace firms like Lockheed and Boeing to set up offices in Russia and strike alliances with local factories. These large foreign corporations proved to be valuable allies, for they came armed with hard currency instead of useless roubles. Their bargaining power enabled them to impose market disciplines on the Russians. But it was a strange trading environment, which at first left the Americans confused. In the Soviet days, factories never worried about profit-and-loss, or even financial control. They just fudged their figures to fit the State's latest five-year plan. Now, with foreigners demanding to know what they could charge for certain goods and services, Russian managers had no idea how to price their products.

Krikalev, the "last Soviet citizen", soon found himself in a world far more alien than anything he had encountered in space—Houston, Texas. Everything at the Johnson Space Center, from the language, the architecture, the social customs and bureaucratic systems, left him feeling more adrift than when he was stranded aboard Mir. Yet he would soon think of the inside of a space shuttle as his normal place of work. Krikalev was among the first participants in the world's next great space project.

A cutaway view of ESA's Columbus module shows that, for all its sophistication, life aboard a space station is essentially like living in a submarine.

The International Space Station was conceived more than two decades ago as a huge orbiting Space Operations Center complete with laboratories, astronomical instruments and an enclosed hangar to prepare robot space probes and human missions to Mars. NASA had not decided exactly what to use it for, but the station was supposed to be ready for anything, just in case. Gradually, however, its designers realized that it could not perform delicate scientific research while also supporting other space missions. Experiments in weightlessness would be disturbed by vibrations from the construction activities in the hangar. Meanwhile, the politicians who were being asked to fund the station argued against the hangar, because they could see no immediate prospect of creating additional spacecraft to be serviced inside it.

President Reagan announced his approval for a slightly less grand version during his State of the Union Address on January 25, 1984. "Tonight, I am directing NASA to develop a permanently manned space station, and to do it within a decade. We want our friends to help us meet this challenge and share in the benefits. NASA will invite other countries to participate so we can strengthen peace, build prosperity and expand freedom for all who share our goals." NASA's engineers still could not agree on the final design. Researchers who needed absolute weightlessness for their experiments argued that the station itself created a tiny gravitational field. They wanted a balanced shape, with the main mass at the center and the rest of the weight evenly distributed to smooth out the gravity field. A new "dual keel" design placed the crew modules in the middle of a rectangular boom structure, with huge solar panel arrays on either side, and this became the station's official layout.

NASA chiefs thought they had finally decided what shape the station should be, but when Bill Clinton became president in 1993 he demanded a cheaper design. However, several giant US aerospace corporations (in particular Boeing and Lockheed) had already committed many thousands of staff to the project. Clinton found it hard to cut costs because so many jobs would have been lost around the country. The crew modules remained largely unchanged, but some savings were made by simplifying the station's complicated boom framework into one central spine. With Clinton's support, NASA chief Dan Goldin and his Russian counterpart Yuri Koptev struck a deal in Moscow to secure Russia's partnership in the project. As they signed the paperwork, vibration from passing tanks shook the windows of Koptev's office. With the station's scientific purpose still undefined, the peace-making benefits of an international collaboration seemed the surest way of justifying the expense.

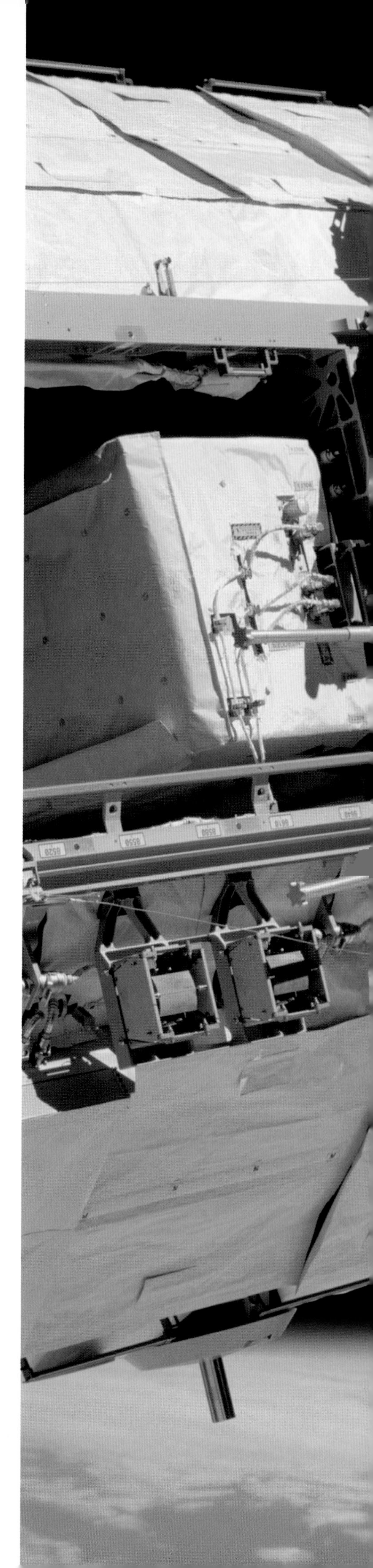

Soichi Noguchi, left, from the Japanese space agency JAXA, works outside the space station with NASA spacewalker Stephen K. Robinson in August 2005.

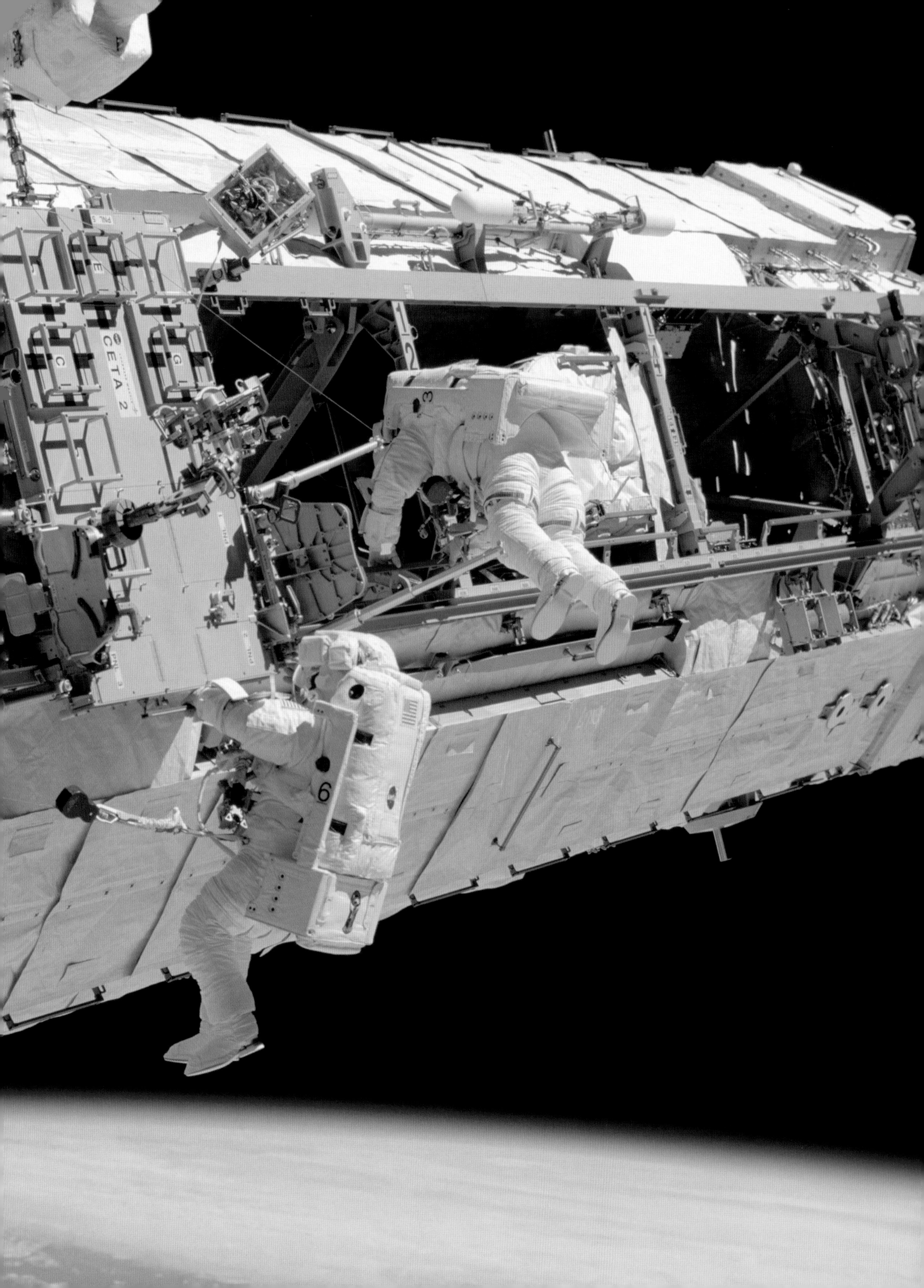
CETA 2

Orbital Culture

> Real-life space stations don't provide their inhabitants with artificial gravity. The gentle spinning of a station was once seen as a way of making life more comfortable for its crew, now total weightlessness is seen as the main justification for building one in the first place. Many science experiments that are impossible on earth can be done in space, precisely because of the weightless conditions. So far, however, the experiments conducted on the International Space Station have not caused as much excitement as has had been hoped in the scientific community on the ground. It's quite a challenge for hard-pressed astronauts to perform delicate science experiments. Maintaining the complicated machinery of the actual station takes up much of their time.

Long-duration missions can create friction between the closest teams. In 1978, a Salyut crew had to be brought home early because one cosmonaut became overstressed. Mir's crews have often bickered, sometimes with each other and more often with mission control, the traditional target for all astronauts' frustrations. In early 1995, prior to the first shuttle visits to Mir, cosmonaut Gennady Strekalov made a number of spacewalks to move solar panels and other equipment on the exterior of the station so that the shuttle would have a clear approach path. As the spacewalk schedules became more intensive and, in Strekalov's opinion, increasingly haphazard, he rebelled. "These spacewalks have not been properly rehearsed. They are dangerous. I won't do them!" Eventually, Strekalov relented and carried out his work flawlessly, but on his return to earth he was fined a substantial proportion of his salary. The American Skylab rebels had not suffered such humiliations on their return to earth.

Occasional tensions between crew members are usually offset by a common bond of training, loyalty to mutual employers and a shared cultural background. Today's multinational space station crews face new difficulties. Norman Thagard, the first American astronaut to spend time aboard Mir, found the cultural isolation hard to bear. The intermittent nature of communication with ground stations meant that sometimes he had to wait two or three days at a stretch before hearing a familiar American voice for a few precious minutes. Thagard had spent many weeks in Star City, the cosmonaut training complex near Moscow, and had learned to handle all the Russian machinery, space suits, valves and airlocks. Once in orbit, the psychological isolation took him completely by surprise. His Russian was good enough to make conversation, but not so good that he could feel close with his crewmates or crack jokes with them. Somehow his training on all the machines had not prepared him to deal with Russian people.

The interior of the space station is not as smart and tidy as the science fiction predictions suggested it should be.

"Space only seems large. For human beings, it is confining. That is why, despite the size of the starry firmament, the idea of space travel gives me claustrophobia."

US journalist and critic Stanley Kaufman, 1968

With these kind of challenges in mind, McDonnell Douglas, the company partially responsible for equipping the interior of the ISS, was the first NASA contractor ever to hire an anthropologist to study the interactions of people in space. Dr Mary Lozano interviewed astronauts from America, Europe, Japan and Russia, and discovered significant cultural preconceptions. "We know people can get on each other's nerves, especially in what we call a 'trapped environment'. Imagine the problems that occur if you come from widely differing cultures."

Dr Lozano's Japanese interviewees, for instance, felt that American astronauts and mission controllers tended towards making rapid judgments without consulting their superiors. NASA's flight staff, often called upon to make split-second decisions that might be crucial to the safety of a mission, thought that the Japanese preference for group decision-making could be fatal in an emergency. Americans felt that the Japanese tended to say "yes" when they meant "no" because of their ingrained desire to maintain social harmony. To the Japanese mind, the Americans' more individualistic approach seemed dangerous and irrational. They thought of Americans as arrogant and thoughtless towards others. Americans considered the Japanese clannish and the French arrogant. Germans thought the Italians over-emotional. Italians found that Americans had little sense of personal privacy. Canadians were annoyed at being thought of by Europeans as "just the same as" Americans.

Dr Lozano also found that Russian cosmonauts occasionally saw NASA's astronauts as too overtly competitive with each other. These particular studies for McDonnell were conducted in 1992-93, but a subsequent incident bore out Lozano's findings. When the shuttle Atlantis returned from the first Mir docking in early July 1995, it brought Norman Thagard home from his three-month stay aboard the station, along with two Russian cosmonauts, Vladimir Dezhurov and the now-contrite Gennady Strekalov. Russian doctors have a routine for bringing crews home. The first rule is that they should not try to walk, or even stand, when they touch down. Couches are brought up to the returned Soyuz and the cosmonauts flop into them for a few hours of rest before struggling to stand. But Thagard wanted to get on his feet as soon as Atlantis had rolled to a halt. For him, it was a personal challenge to overcome stiff joints and dizziness and prove that he could walk despite his long sojourn in weightlessness. His Mir companions thought he was trying to show them up as inferiors. That certainly was not Thagard's intention. It was an accidental clash of cultures.

McDonnell's inquisitive anthropologist discovered that Dutch and French members of the European Space Agency's Spacelab program had an altogether more pressing problem on their minds. They worried that

A pair of Russian spacesuits, designed for use aboard a Soyuz capsule, wait for their owners to climb into them.

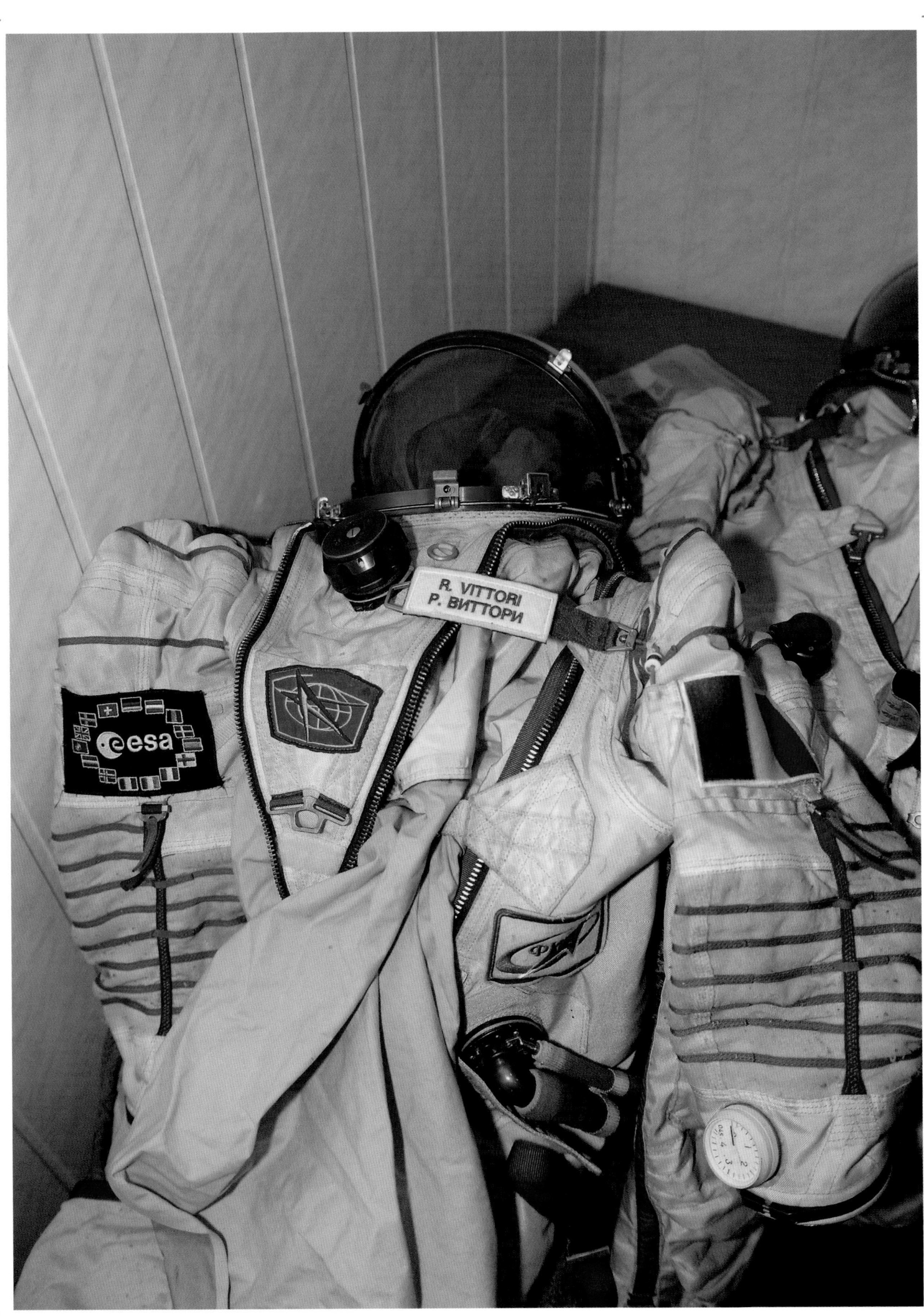
R. VITTORI
Р. ВИТТОРИ
esa

mealtimes were not taken seriously enough aboard NASA shuttles. They liked good food and treated dinner time as a social break. American crew members seemed to eat just because it was time to refuel their bodies. Minor attitudes toward personal hygiene and body odor also emerged as an issue.

Instrument layouts proved another area rich in cultural ambiguities. Britons may "switch" systems on, while Americans sometimes "turn" them on (a legacy of the familiar design of domestic light fittings). Meanwhile, the American custom of flicking a switch upwards for "on" is not universally recognized by other nations, and this distinction can be critical in space. Flashing red or yellow warning lights inform NASA crews that something is wrong. To the Far Eastern mind, red can symbolize good fortune and so its use on instrument panels does not always conform to western ideas, and so on.

Then there are subtleties of language to do with flight-deck procedures. No matter how accurate the translation of some term might be, its meaning can be radically different. Americans often use the word "control" for switches, levers and dials: "thrust controls, attitude controller" and so on. To Russians, this word "controller" has a deeply ingrained administrative association. Ask cosmonauts to locate the attitude controller and they might think in terms of some superior officer on the ground who authorizes which attitude they should fly in. It does not help that NASA astronauts also respond to instructions from their "mission controllers" in Houston to adjust some particular "controller" on their spacecraft. International training regimes rely on finding a common new language to try and eliminate ancient confusions.
In fact, these challenges of forging a cooperative environment among such different cultures is one of the main benefits of the space station, above and

Opposite, a Soyuz booster is rolled out of its construction hangar. Russian industry is adept at turning out boosters, but foreign investment is essential to keep the production lines moving.
Above, Russian cosmonaut Aleksandr Kaleri, and the mission commander (top), are followed by British-born NASA astronaut Michael Foale and ESA astronaut Pedro Duque from Spain, as they prepare to climb aboard their Soyuz spacecraft for a flight to the International Space Station on October 18, 2003.

beyond its fitful science programs and uncertain role in future deep-space exploration. This is the most complex and expensive international collaboration ever attempted in peacetime history, and although the station may not be able to "pay its way" in any immediate practical sense, it does have a profound value to us. Perhaps the best reason for building it is simply to prove that we can, and that so many nations can work together on a common project. If the station serves as an example of global collaboration to inspire future generations, then its $100 billion costs will have been worthwhile.

The slender tube on the very tip of the assembly is the launch escape rocket. This pulls the Soyuz capsule clear of the main booster in the event of any launch failure. A similar escape rocket saved the lives of cosmonauts Vladimir Titov and Gennady Strekalov in September 1983, when their booster exploded on the pad.

Above, technicians at the Yuri Gagarin cosmonaut training center at Star City, near Moscow, prepare a cosmonaut for underwater training, to simulate a spacewalk.
Opposite, a Soyuz rocket lifts off from the pad at Baikonur, Kazakhstan, carrying three humans into space. What was once seen as something amazing is now regarded as routine.

07 WHAT NOW, VOYAGER?

"The government is forcing the space program to be done with technology that we already know works. They are not creating an environment where it is possible to have a breakthrough."
SpaceShipOne designer Burt Rutan, 2006

WHAT NOW, VOYAGER?

If more space stations are to be built in the future, they might well take new forms that NASA has not yet seriously considered. They could become places for ordinary people to visit, and perhaps even to call their home. Some theorists have imagined stations as large as towns, serving social rather than scientific needs. Huge populations would live in orbit, far above the earth. In 1903 the Russian pioneer of astronautics, Konstantin Tsiolkovsky, (who worked on detailed ideas about rockets from the 1880s until his death in 1935), proposed a huge habitable cylinder spinning on its axis and containing a greenhouse with a self-supporting ecological system. And in the novel *The World, the Flesh and the Devil* (1929) J.D. Bernal envisaged "Worldships" capable of housing thousands of people. These colossal concepts were revived in the 1970s by Gerard K. O'Neill, a physics professor at Princeton University. In the wake of the Apollo lunar landing of 1969, he proposed huge colonies in space. The structures would be perhaps 3 miles in length, fabricated from materials processed in lunar factories. At first his ideas were intended mainly as a theoretical exercise to stretch the imaginations of his physics students, but the concept struck a chord during the 1970s with its studies of the "closed" ecological systems that might be required to support a large population inside a finite little universe. Environmental campaigners on earth found some kindred spirits among the space colony crowd. Especially popular was O'Neill's proposal that all heavy industry and energy production should be taken into space to save the earth from the burdens of pollution. Unfortunately, current rocket launch technology is too expensive and fuel-intensive to enable O'Neill's vision in our lifetimes. Any large-scale space colonization project would harm

The success of SpaceShipOne, a product of Burt Rutan's Scaled Composites company, has encouraged other private industrialists to design new machines for space.

Left, in the 1970s there was much talk of building giant space colonies housing thousands of people in communities the size of cities.
Opposite, South African internet entrepreneur Mark Shuttleworth trains aboard a Soyuz simulator prior to his flight in 2002. His experiences in space inspired him to set up education initiatives for his home country.

the environment rather than sparing it. A tremendous number of rocket launches would be required to get it off the ground.

Such a scheme would need a genuinely efficient, low-cost, safe and completely reusable spaceship—and one that burns only clean fuel, such as hydrogen. This is a tall order, and so far, none of the aerospace corporations who build today's rocket boosters have been able to finance the successful development of such a machine. The most profitable passengers for space are communications satellites, not people. It's just not worth building complicated reusable spaceships to launch little com-sats. Throwaway launch vehicles are cheaper to develop than genuine "ships" suitable for routine spaceflight. The basic design of most rocket boosters has changed hardly at all over the past fifty years. NASA and its fellow space agencies around the world were once seen as the vanguards of a new technology. Now they may be in danger of becoming the custodians of an old one.

Could space tourism give the rocket industry a vital injection of modernity? In 1994 the American Society of Civil Engineers held a conference in Albuquerque, New Mexico, to talk about holidays in space. Beautiful models of orbiting hotels and passenger-carrying shuttles were presented. Many delegates were impressed by the technological ideas on display. Just as many were persuaded that it was an impossible dream. Financial experts had advised them that ordinary people would not be willing to pay the extraordinary amounts of cash required to visit space.

The pessimists spoke too soon. There are indeed people willing to pay for a taste of space. In 1998, the Virginia-based company Space Adventures negotiated with Russia's Star City cosmonaut training complex to provide holidaymakers with $10,000 weekends of dressing up in spacesuits and clambering aboard simulator capsules. Customers also signed up for zero-gee training flights aboard "vomit comets", jet cargo planes flying a high parabolic arc so that passengers can experience a couple of minutes of weightlessness at the top of the arc.

This was just the more affordable end of the Space Adventures menu. On April 28, 2001, American financier Dennis Tito was sealed inside a Soyuz capsule as he began the world's most expensive holiday, accompanied by two Russian cosmonauts on their way for a tour of duty aboard the International Space Station. Tito had committed $20 million of his own cash to secure this opportunity and booked his flight through Space Adventures, with cooperation from the Russian Space Agency. NASA was not keen on the flight and so Tito had to remain inside the Russian-built modules of the station throughout his week-long visit. Certainly his Russian hosts made him feel welcome. Cosmonaut Yuri Baturin told him: "We are very happy to accompany you to space. We like your mathematical mind and we like even more your romantic soul."

NASA acknowledged that the cash from this unusual mission had been important for Russia's space industry. A year later, South African-born internet

"When once you have tasted flight, you will forever walk the earth with your eyes turned skyward."
Private space explorer Mark Shuttleworth, 2002 quoting Leonardo da Vinci

"I hope, with the launch of Virgin Galactic and the building of our fleet of spacecraft, that someday, children around the world will wonder why we ever thought space travel was just a dream we read about in books or watched, with longing, in Hollywood movies. If we can make space fun, the rest will follow. This is a business that has no limits."
Richard Branson, 2005

Above, SpaceShipOne taxis for a test flight, slung underneath its carrier aircraft, White Knight. Opposite, the XCORP company's proposal for a small suborbital passenger craft that will take off and land like a normal jet plane.

entrepreneur Mark Shuttleworth made a similar voyage and the rules concerning access to other areas of the station were relaxed. On his return to earth, Shuttleworth set up a foundation to encourage the teaching of math and science in Africa.

So far at least, space tourists have shown as much dedication to their missions as any space professional. In October 2005, when Dr. Greg Olsen became the third private entrepreneur to visit the station, he was mildly disturbed to be thought of as a mere tourist. "I spent over nine hundred hours in training. There were many exams, medical and physical, as well as classroom and competency tests. It's not like you pay your money and go on a ride. You have to qualify for this." As the founder of a New Jersey company specializing in optical and infrared sensors, he conducted serious scientific experiments during his time aboard the station. He took care, however, not to compare his qualities with those of a full-time space professional. "I don't consider myself an astronaut. I'm a space traveler, and I've been in orbit. But I have far too much respect for astronauts and cosmonauts to call myself that."

For those who have made a fortune on the stock markets or the internet, $20 million in travel expenses may not be impossible to find, but no one expects many repeat performances at that price. The future of space tourism will rely on bringing ticket costs down to thousands rather than mllions of dollars. In 2002, market researchers Zogby International surveyed a cross-section of American business leaders about space tourism. One in five confirmed they would be happy to pay $100,000 for a suborbital flight, and seven out of every hundred in the super-rich category said they would spend a million dollars or more to secure an entire week in space, although Tito and Shuttleworth's investment of $20 million could be matched by only a very few individuals.

Space tourism will make little further progress without a genuinely cost-efficient space liner. In May 1996 a group of business people held a gala dinner in St. Louis, Missouri, to celebrate Charles Lindbergh's historic 1927 first flight across the Atlantic, funded by an earlier generation of patrons from that city. They were competing back then for a $25,000 prize set up by New York hotelier Raymond Orteig. That was a substantial sum of money in 1927. Could a similar prize spur modern private industry to develop cheap, affordable human spaceflight? And so the $10 million Ansari X-Prize was created, under the leadership of Space Adventures co-founder Peter Diamandis, to reward the first private reusable passenger-carrying ship to reach the edge of space.

SpaceShipOne and White Knight take off from Mojave Airport in California, home base for the Scaled Composites company.

WHITE KNIGHT

"The genie is out of the bottle, the fuse has been lit. We are really at the birth of the personal space flight revolution."
X-Prize Chairman Peter Diamandis, February 2005

Anousheh Ansari couldn't speak English when she immigrated to the United States from Iran in 1984 at the age of 16, but she hoped that living there would help her realize her dream of becoming an astronaut. She enrolled at George Mason University to study electrical engineering. There she met Amir and his brother Hamid, whom she married in 1991. The three of them eventually founded their own telecommunications company. By 1996, Telecom Technologies Inc. was the fifth fastest-growing technology company in Dallas. Anousheh never lost her fascination for the human adventure of spaceflight, and she and her company were major sponsors of the Ansari X-Prize. "As a child I looked at the stars and dreamed of being able to travel into space," she said. "As an adult, I understand that the only way this dream will become a reality is with the participation of private industry and the creative passion of smart entrepreneurs."

A dozen companies took up Ansari's challenge. A fascinating array of designs included delta-winged space planes, eerie capsules shaped like seed pods, and even a machine with rocket-powered helicopter blades. Some competitors built real hardware, while others were unable to finance much beyond computer-generated concept artworks or unpowered mock-ups. But the seeds for a new industry were sown, and a number of companies that did not make the cut first time around are still in business: older, wiser, less naive about the financial burdens of research and development, yet undaunted.

Burt Rutan's elegant SpaceShipOne eventually claimed the prize on October 4, 2004 (the choice of Sputnik's anniversary date for the flight was deliberate). The project immediately won back half the $20 million invested by Microsoft's adventurous co-founder Paul Allen. And just as Lindbergh's flight inspired a huge transatlantic air industry, Rutan's team may have triggered a similar gold rush. Two days after the final qualifying flight, British entrepreneur Richard Branson announced that his Virgin Group had already pledged another $25 million towards SpaceShipOne's successor, along with supporting ground facilities at a dedicated site in New Mexico. An additional $100 million was pledged for building a small fleet of "Virgin Galactic" suborbital space liners, each with half a dozen passenger seats plus two at the front for the pilots. For just under $200,000 customers will experience a rocket flight into the vacuum of space. They will have a few minutes to enjoy a stunning view of the earth's curving horizon before their little ship plunges back toward the ground.

More than 13,000 would-be space voyagers from around the world have already expressed a desire to pay a deposit with Virgin. They range from anonymous business people and private individuals to glamorous Hollywood stars. Some want to pay the entire ticket

Artist Pat Rawlings depicts a lunar passenger ferry leaving earthy orbit for its two-day journey to our neighboring world.

LUNOX

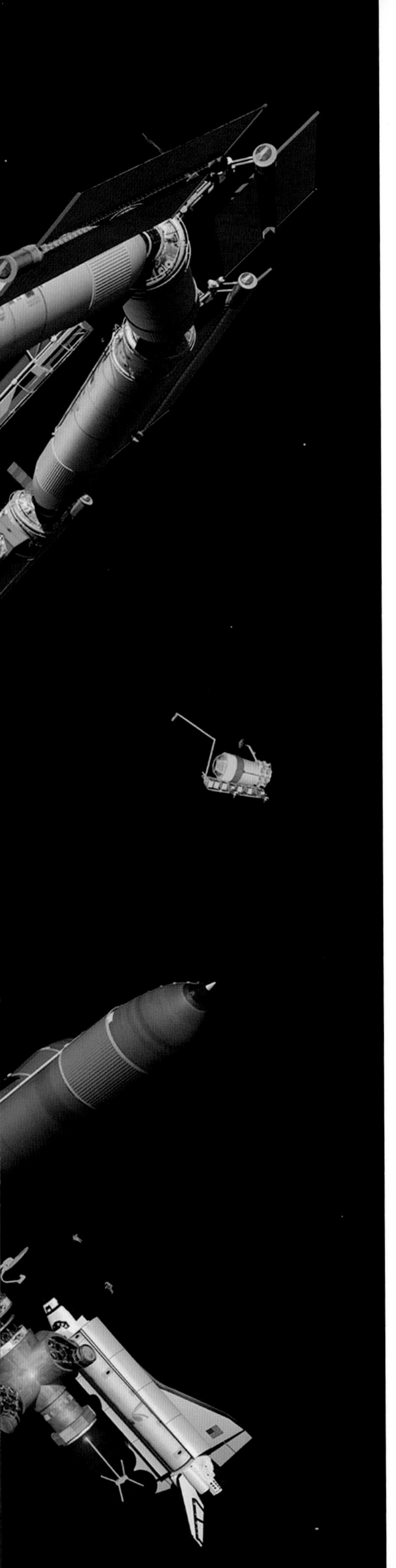

A rotating space station assembled from empty space shuttle fuel tanks.

price immediately, guaranteeing themselves a seat on one of the first flights. Virgin Galactic does not expect to fly more than one mission a week during its somewhat experimental first few years of launch operations. The company is determined not to hurry the maintenance and turn-around times of their vehicles between one take-off and the next. Even so, with bookings worth $1 million per flight, Virgin Galactic could be in profit after just four years of launch operations.

For Branson, space is a business opportunity that also satisfies a personal yearning for adventure. "I hope, with the launch of Virgin Galactic and the building of our fleet of spacecraft, that someday children around the world will wonder why we ever thought space travel was just a dream we read about in books or watched, with longing, in Hollywood movies," he explained when he announced his new company to a startled world. "If we can make space fun, the rest will follow. This is a business that has no limits."

While Rutan's Scaled Composites company has received the first Federal Aviation Authority license for a private manned suborbital spacecraft, several other companies are developing suborbital vehicles. XCOR Aerospace is among the serious contenders, and Blue Origin, a Seattle company backed by Amazon.com founder Jeff Bezos, has the resources for a major project to rival Virgin Galactic.

If $200,000 a seat is still too expensive for most people, it may soon become possible to experience

NDD
NEXT DAY DELIVERY
EPS
Eagle Parcel Service

Opposite, a commercial lunar landing ferry delivers people and cargo to the surface, some time in the next 30 years and right, astronauts survey permanent constructions on the surface.

space for little more than the price of a luxury automobile. In Japan, where the idea of space tourism seems to have great popular appeal, the Kawasaki company has been studying plans for Kankoh-Maru, a fifty-seat ship shaped like a giant lozenge with a flat base, which could be developed in the next decade or so. Kawasaki believes that passengers could be asked to pay as little as $50,000 for a seat. This could still create revenues of $2.5 million per flight.

But getting off the ground is only half the story. There is a limit to how far space tourism can progress when there is nothing for the passenger ship to do except fly up and down. And this is where a new generation of privately financed space stations may find a role. Las Vegas businessman Robert Bigelow, founder of the Budget Suites of America hotel chain, wants to create the first orbiting space hotel. He will use "TransHabs"—inflatable modules already developed by NASA but not yet flown into space.

First, Bigelow needs to find a cost-effective courtesy shuttle to deliver customers to his hotel and bring them home again. To make this happen, he has taken the X-Prize idea one step further by offering a $50 million reward for the first company to fly passengers all the way into orbit aboard a reusable ship. This is a much greater challenge for a spacecraft than merely dipping

The crew of the lost shuttle Columbia. From left to right: David M. Brown, Rick D. Husband, Laurel B. Clark, Kalpana Chawla, Michael P. Anderson, William C. McCool, and Ilan Ramon.

its nose above the atmosphere for a few minutes, before plunging back to earth, but this challenge must be met if space hotels are to become a reality.

Dennis Tito certainly hopes that people less wealthy than him will soon have a chance to share his incredible experiences. "I think one of the things that will keep people interested in space is seeing other ordinary people go up, from all walks of life—people they can identify with."

The main doubt clouding this bright new horizon is the sheer number of flights that will be needed to make space tourism pay, and how those flights will add to the noise and pollution already dumped by the aviation industry into our fragile atmosphere. Other legal and safety aspects of this new industry are the subject of complex debates in the U.S. Congress. The Federal Aviation Authority (FAA) controls safety over American skies. No civilian machine can leave the ground without an FAA license. In 2004 Congress decided the FAA should also regulate the carrying of passengers into space. The FAA at first demanded that orbital holidaymakers should be as safe in a rocket ship as in a normal plane. No one could guarantee that and it seemed as if the space tourism industry might be stifled at birth by over-cautious legislation.

In December 2004, Congress ruled that space passengers must simply accept some of the risks. They cannot automatically expect compensation if early experimental flights prove hazardous. The FAA agreed to relax its rules for the next eight years, allowing serious space businesses—operating in good faith—to build up a flight history. When passengers eventually queue up in larger numbers it will be time to think about safety in more detail. Eventually the FAA could claim a substantial reward for its patience: a powerful new office dedicated to safety regulation and traffic control in space as well as the air.

While private industry looks forward to the commercial possibilities of orbital habitats, government-funded space stations as we know them today have yet to find their true purpose. Half a century has passed since von Braun's famous series of articles in *Collier's* magazine, and the station concept still hasn't proved itself absolutely essential to our continued scientific exploration of deep space. At the same time, the American public is disenchanted with the fact that NASA's astronauts never venture further than a few hundred miles up into the sky.

The loss of a second shuttle in 2003 presented a grim opportunity for reflection and reinvention. A year later, President George W. Bush pledged that America's primary contributions to the International Space Station must be phased out by 2010, at which point the complex and dangerous shuttles that carry modules and crews will be retired forever. A new generation of spacecraft will then reach for a more distant target: the moon.

Back to the moon

> On February 1, 2003, space shuttle Columbia disintegrated during reentry. The entire crew of seven astronauts was lost. An investigation over the ensuing months concluded that a suitcase-sized chunk of foam insulation from the huge external fuel tank had peeled off during launch. It slammed into the front of the shuttle's left wing, making a small but ultimately lethal hole in the heat-resistant panels. This disaster, horribly echoing the Challenger disaster of 1986, forced America to reconsider its entire space program.

The Columbia Accident Investigation Board (CAIB) concluded that it is too risky to carry astronauts in the same part of a spacecraft that contains the rocket engines, because of the risk of explosions, collisions or aerodynamic break-up of the structure. The Board noted that the 1960s Apollo capsules were remarkably safe overall. The tough, compact crew compartment could always be instantly separated from other modules or booster rockets. When a section of the Apollo 13 spacecraft exploded on the way to the moon in April 1970, the rear service module with the rocket engine was torn open, but the crew capsule itself was unharmed and eventually returned safely to earth. The shuttle has no such self-contained "command module" capable of cleanly separating its crew from the rest of the machinery in the event of a problem.

In NASA's latest plan to re-initiate lunar exploration over the coming decade, a launch vehicle created from an adapted space shuttle main fuel tank and solid rocket boosters delivers a lunar landing craft to earth orbit.

After docking with a Crew Exploration Vehicle (CEV) the lunar vehicle is blasted towards the moon.

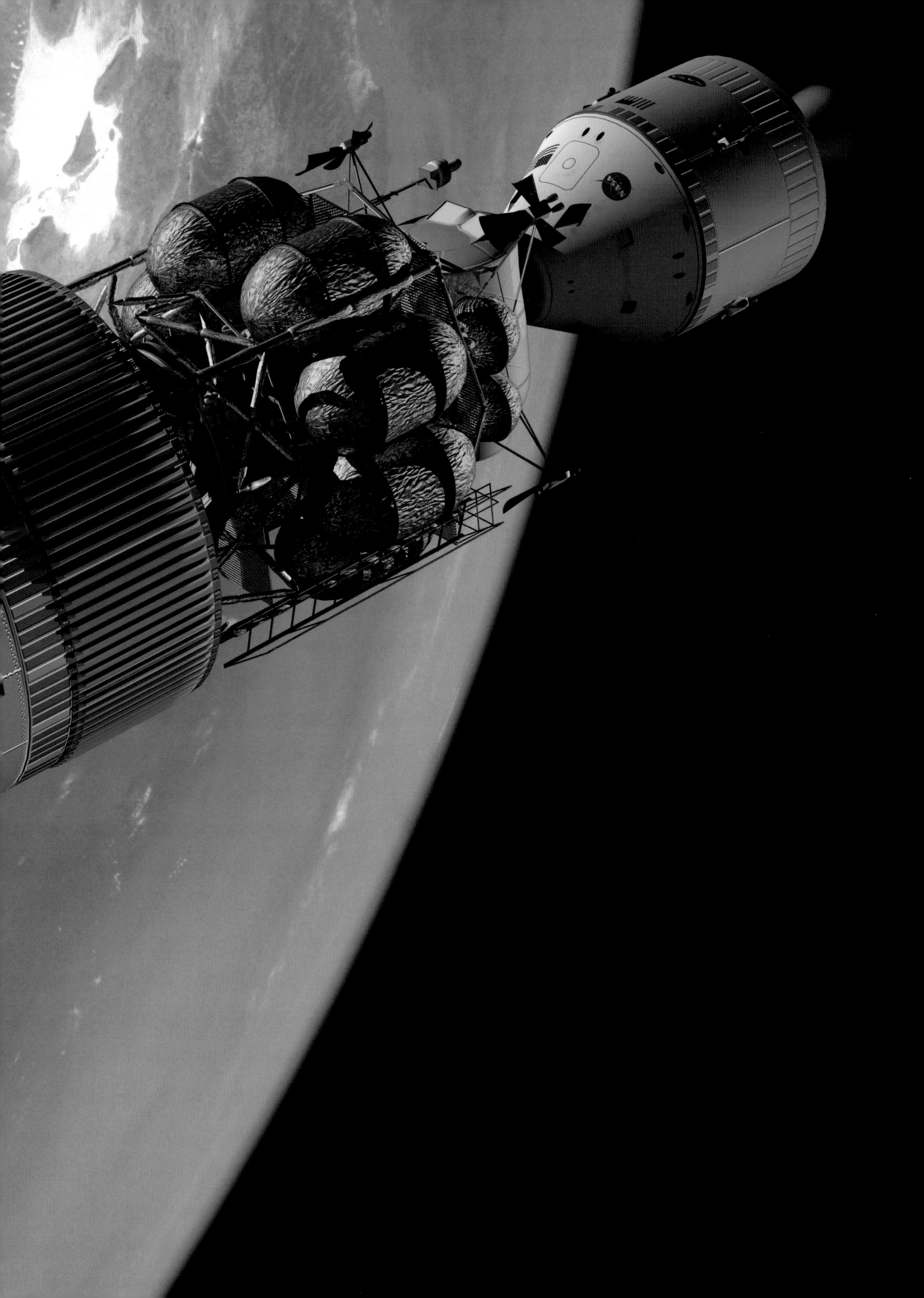

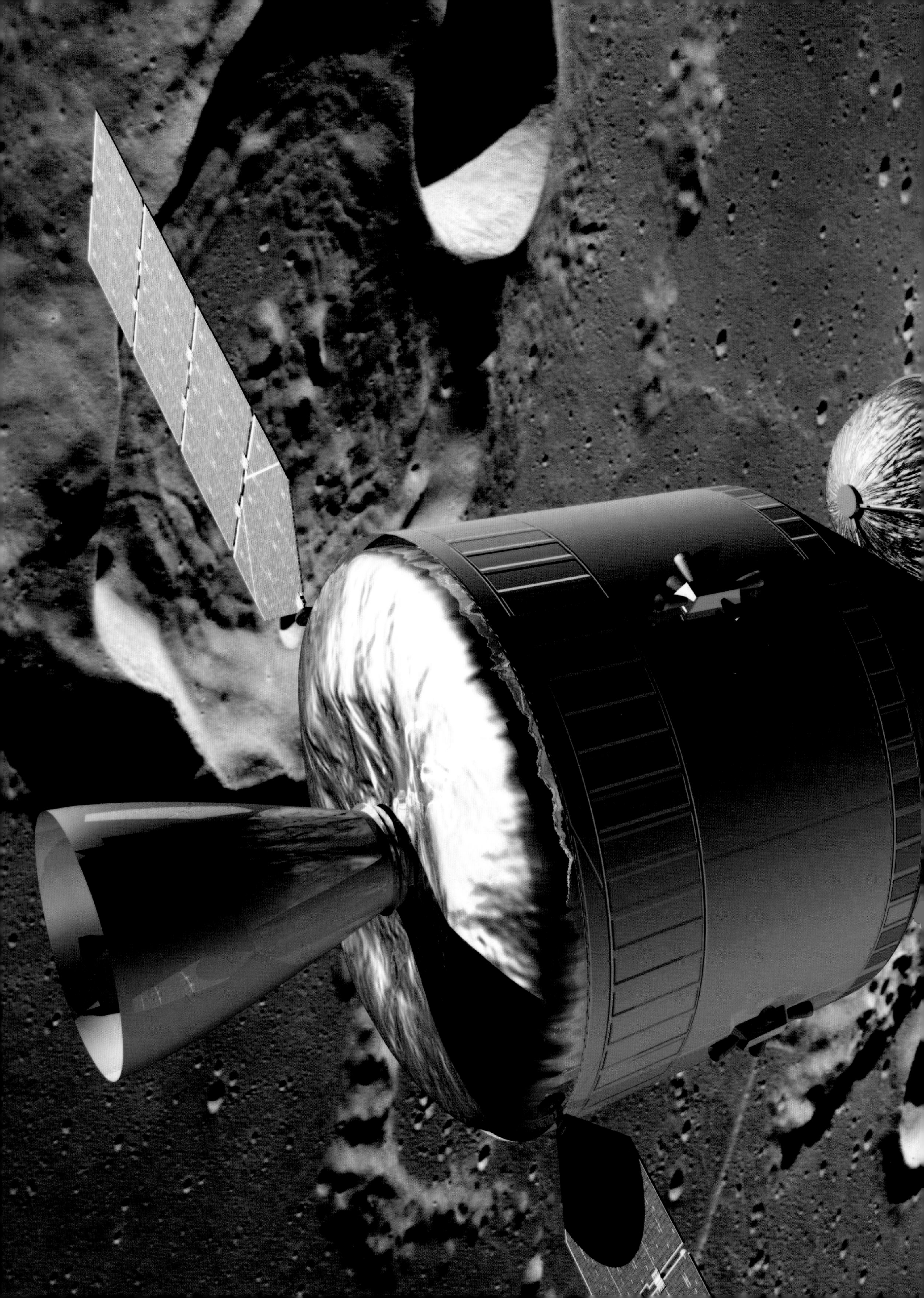

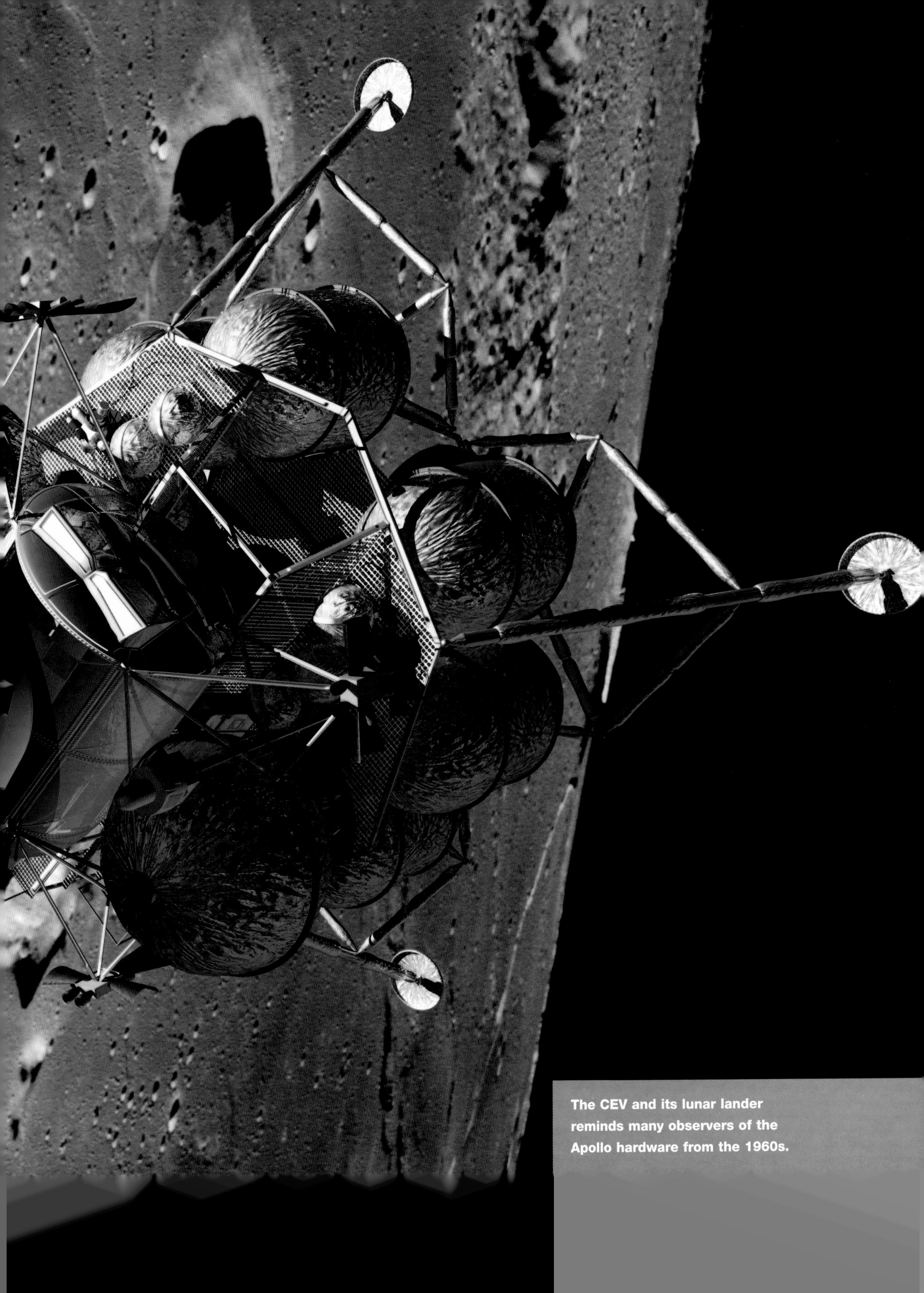

The CEV and its lunar lander reminds many observers of the Apollo hardware from the 1960s.

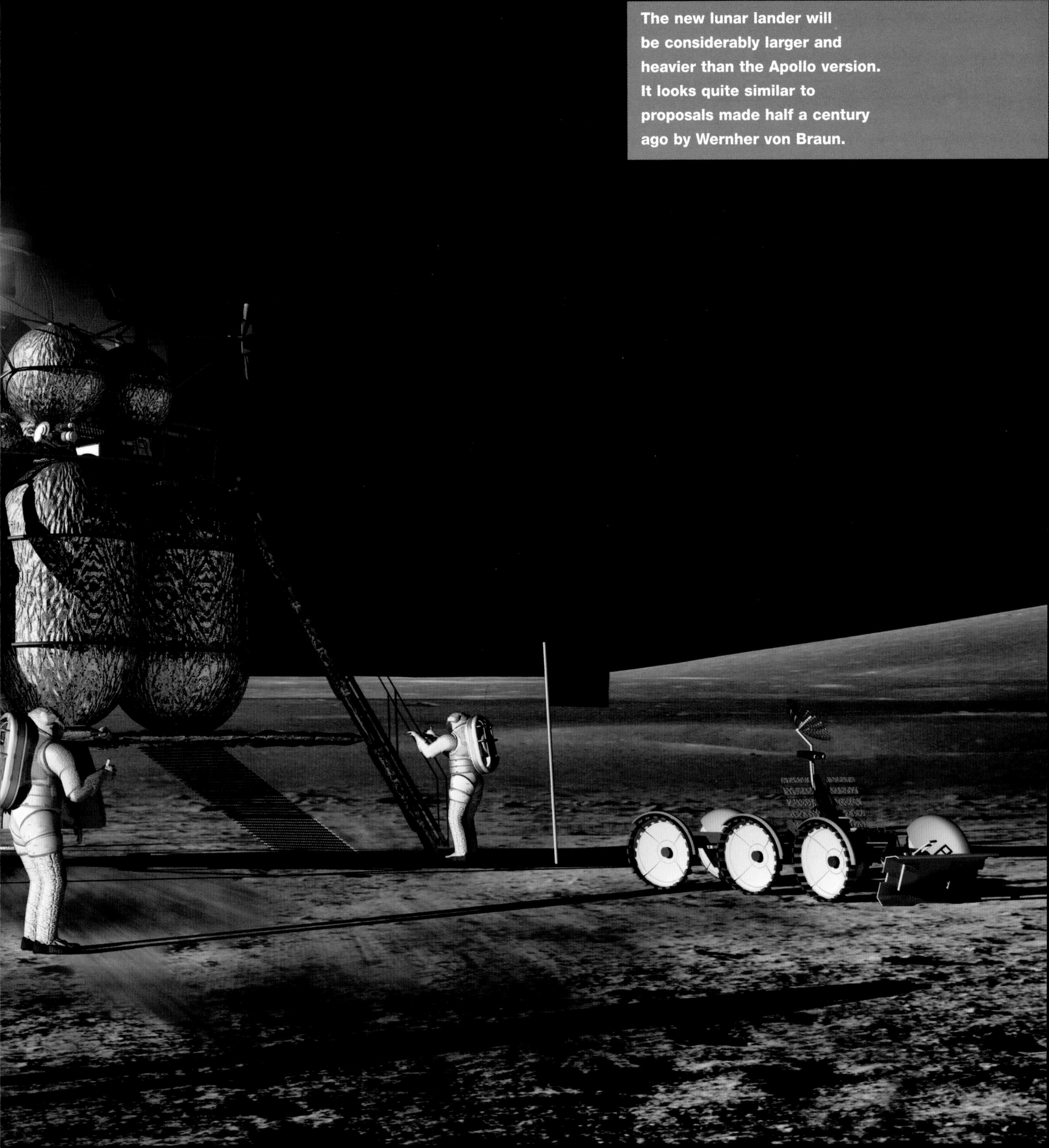

The new lunar lander will be considerably larger and heavier than the Apollo version. It looks quite similar to proposals made half a century ago by Wernher von Braun.

Above, is the moon our destined new home, or will we always be strangers there? Does America have the political will to realize its new vision for space exploration?
Opposite, a new generation of reusable lunar suits with toughened exteriors will be developed in the coming decade.

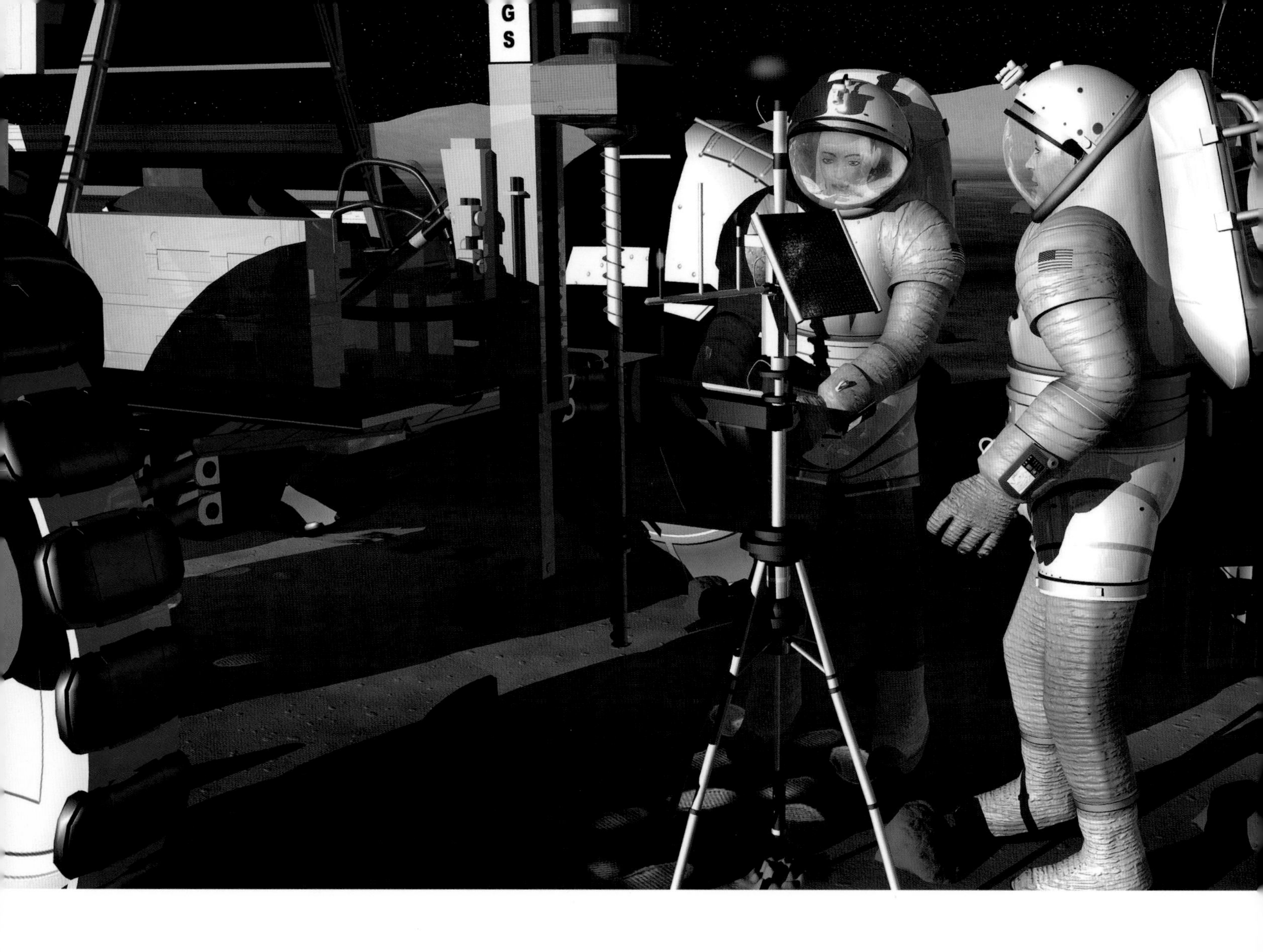

"If we choose to travel into space in the future, it should not just be to conquer new worlds and replicate our sprawling shopping malls on other planets. We should go out there to discover more about ourselves—and to look back, with wiser and more appreciative eyes, at the world we leave behind."
Apollo 8 astronaut Bill Anders

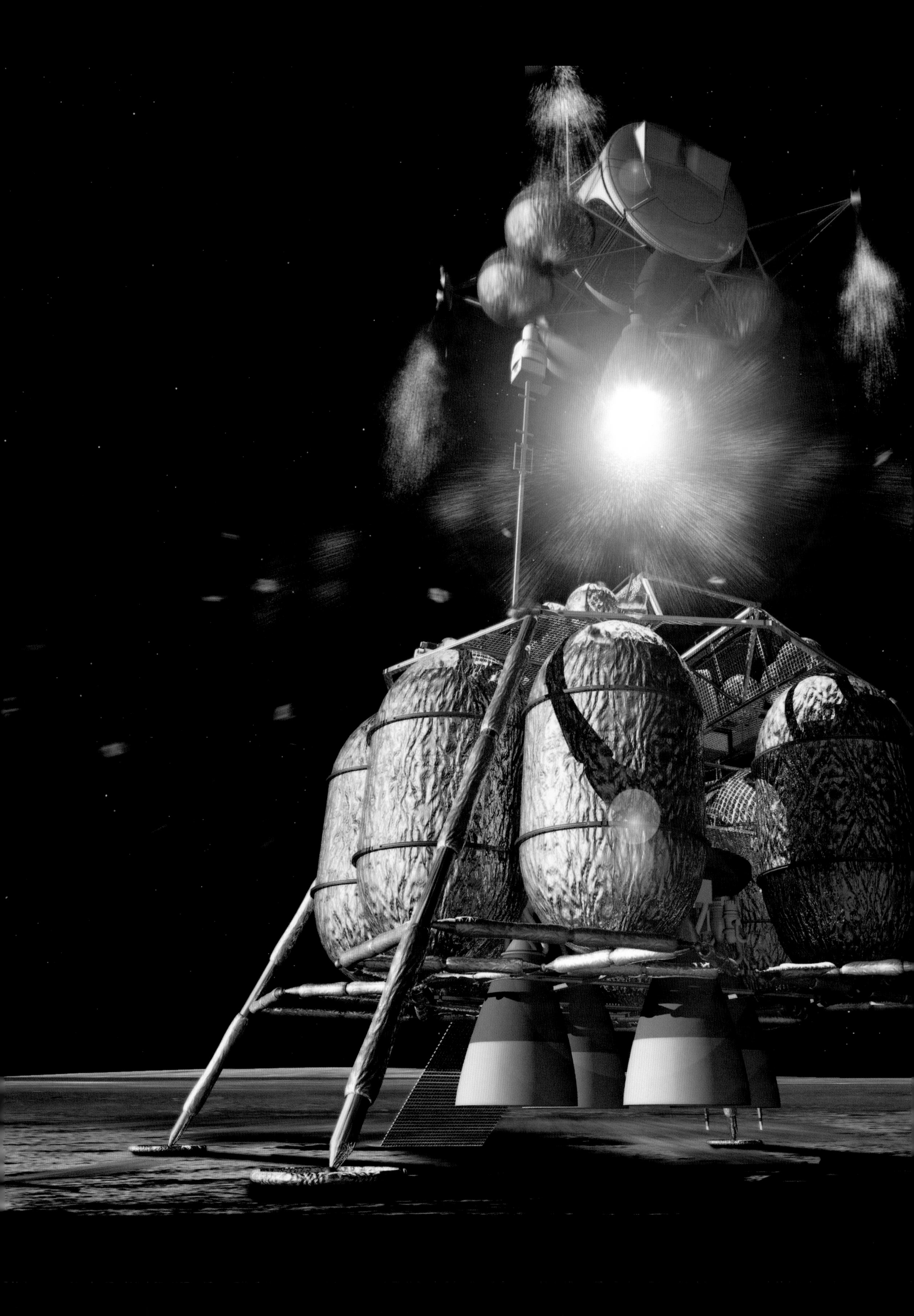

The ascent stage of a lunar lander fires its crew back into orbit at the start of their return journey to earth. Will our next mission to the moon be just another brief visit, or a permanent colonization?

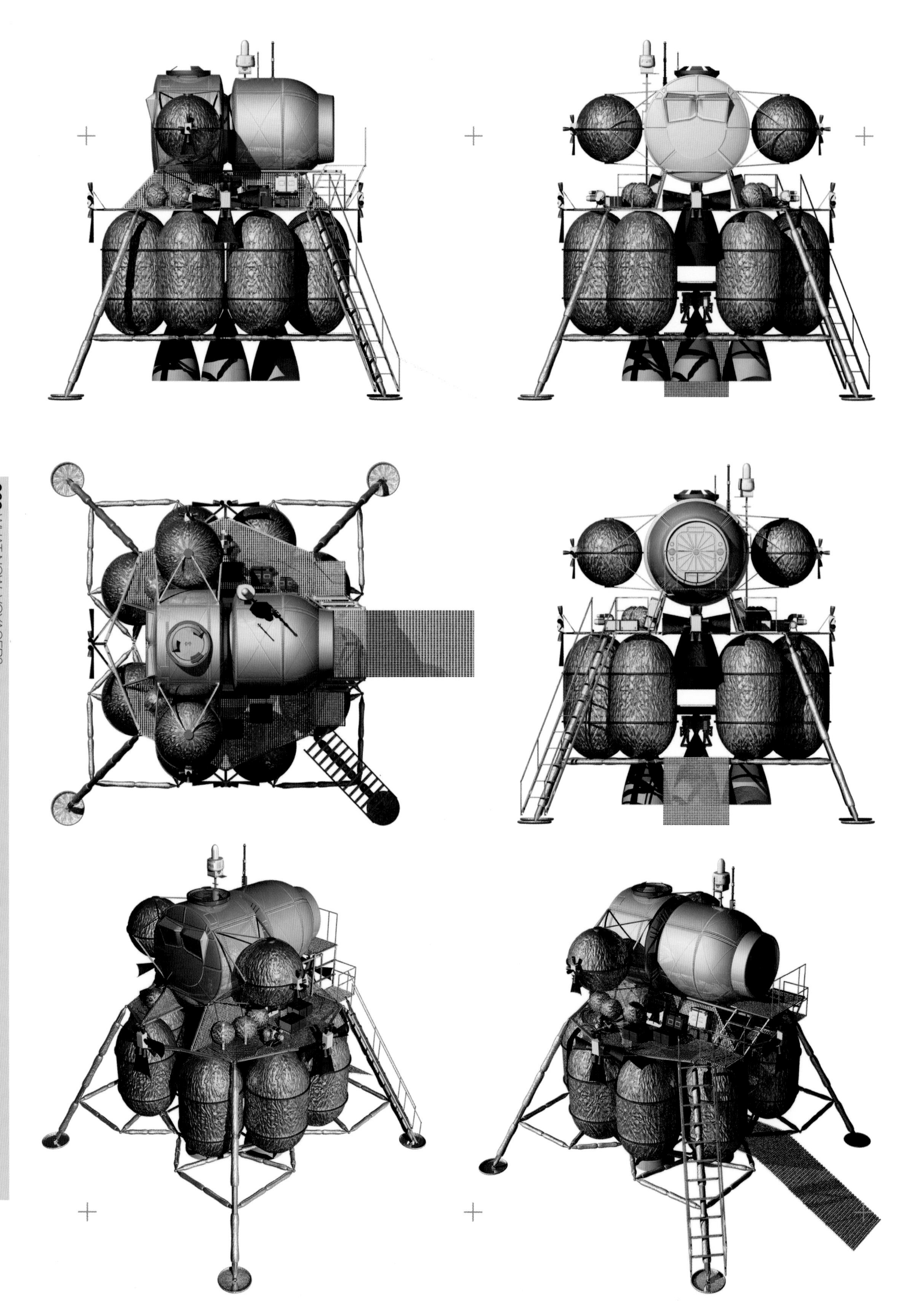

"The idea of establishing a permanent, and eventually self-supporting, colony on the moon will seem to many an even more fantastic idea than that of crossing space itself. Yet, on sober analysis, there is nothing impossible about it—and if there is one thing that recent years have shown us, it is that anything that is possible is sooner or later done."
Arthur C. Clarke, *Exploration of the Moon*, 1954

The CAIB also recommended that future crew modules for any spacecraft should be carried on the uppermost tips of their carrier rockets, so that no launch debris can fall onto them from hardware above, as had happened in the Columbia disaster (and to a less destructive yet still alarming degree, during many other shuttle missions too). The CAIB also urged that NASA's remaining shuttles, Atlantis, Endeavor and Discovery, should be retired from service by 2010.

But what should replace them? After deliberating with his advisors for many months, President George W. Bush made a televised speech in January 2004 during a special visit to NASA headquarters in Washington.

"Today I announce a new plan to explore space and extend a human presence across our solar system." The first goal, he said, was to return the shuttles to flight status so that they could participate in the long-delayed completion of the International Space Station. No surprises there. But as his speech continued, radical new ideas emerged. "Our second goal is to develop and test a new spacecraft, the Crew Exploration Vehicle, by 2008, and to conduct the first manned mission no later than 2014... Our third goal is to return to the moon... Using the Crew Exploration Vehicle, we will undertake extended human missions to the moon as early as 2015, with the goal of living and working there... With the experience and knowledge gained on the moon, we will then be ready to take the next steps of space exploration: human missions to Mars and to worlds beyond."

Engineering visualizations of the proposed new lunar lander.

"The question of taking on Mars as an interplanetary goal is not simply one of aerospace accomplishment, but one of reaffirming the pioneering character of our society."
Robert Zubrin, *The Case for Mars*, 1996

These plans emerged after months of secret consultation between White House officials and space experts during 2003. Most politicians agreed that the loss of Columbia was a wake-up call. NASA needed a new sense of purpose and Mars was an obvious long-term goal for human spaceflight. But Bush had personal reasons for being cautious. Back in 1989, his father also promised a mission to Mars. NASA presented an overblown $400 billion budget that appalled everyone and the idea was scrapped almost immediately. The son did not wish to repeat his father's mistake, and he made only vague references to a Mars mission in his speech. When it came to the moon, he was more confident about possible dates. After all, NASA already knows how to get there, and it's only a three-day flight away.

Bush's timing was influenced by the presidential elections coming up in November. The 9/11 crisis and the wars in Afghanistan and Iraq had defined his first term as president to the exclusion of almost everything else. He must have welcomed a chance to change the subject. "We choose to explore space because doing so improves our lives and lifts our national spirit," he said. His speech was not especially poetic, but he may have been right when he claimed: "In the past 30 years, no human being has set foot on another world, or ventured farther upward into space than 386 miles, roughly the distance from Washington, D.C. to Boston, Massachusetts.

America has not developed a new vehicle to advance human exploration in space in nearly a quarter-century. It is time for America to take the next steps." The hardware for that next step will be designed soon enough. The greater challenge will be to find the money. Much of it will come from diverting NASA funds currently committed to the shuttle program.

If Bush's vision is to be realized, NASA's contributions to the space station will have to be cut back after 2010, at least as far as major crew transfers and shuttle-carried modules are concerned. The Russian, Canadian, European and Japanese space agencies will assume greater control. They will need to find alternatives to the shuttle as a carrier for their astronauts and supplies. Non-American participants in the station have had to adapt over the past two decades to NASA's shifting fortunes, and this has not been easy. The impetus, now, is on developing an international range of launch vehicles so that each space agency can collaborate without becoming stranded if another's vehicles encounter problems.

The Crew Exploration Vehicle (CEV) is a reversal of much that NASA has worked towards over the past 30 years. The dream of an "all-purpose" shuttle, flying cheaply and regularly like a commercial cargo plane, never came to fruition. Whenever a shuttle puts a satellite or a space station module into orbit, the costs are huge because astronauts always come along for the ride. The launch weight given up to life support and crew cabins limits the amount of useful payload a shuttle can carry in its cargo bay. It would be more efficient, say the CEV's designers, to haul cargo in uncrewed rockets, and launch astronauts separately in

"The public may well be more risk adverse than the people in the space business, but limiting the space program in the name of eliminating the risk is no virtue."
Neil Armstrong, March 2004

a small module that carries people and nothing else. There's no need for the shuttle's huge wings and cargo bay doors. All that's left to worry about is the crew cabin at the front. Replace that with a cone-shaped capsule that fits on the end of a conventional rocket and it can also serve as an escape pod if anything goes wrong during launch: a non-negotiable CAIB requirement.

But is this progress, or a strange leap back in time? Aerospace engineer and writer Robert Zubrin has worked tirelessly over the past decade to promote a human Mars mission. (His lobbying group, the Mars Society, has some 8000 members.) The wayward logic of space planners often exasperates him. "In the 1970s, when President Nixon killed the Apollo program and ended lunar exploration, NASA said we would do all that again one day, after we had developed cheaper transportation to orbit using a winged shuttle. We've spent twenty-five years trying that, with no positive results. The shuttle costs more to fly than the Saturn boosters we had for the Apollo missions. Now the shuttle is going to be replaced with a CEV capsule just like Apollo."

After more than three decades, it is impossible to recreate Apollo's giant Saturn rockets and launch gantries. All the factory tooling was scrapped and even the original drawings have been lost. Instead, the CEV capsule will be launched on top of a solid rocket booster, as currently employed by the space shuttle. Larger components of a lunar spacecraft will be lifted by a Delta IV satellite launcher as unmanned cargo and docked together in earth orbit to form a larger combined spacecraft for trips to the moon. This is an eerie resurrection of concepts championed by Wernher von Braun and Arthur C. Clarke in the 1950s.

On completion of the outward journey to the moon, a landing craft will detach and drop down to the surface. At the end of its mission the lower part of the craft will stay behind while the upper crew module blasts back into lunar orbit, makes a rendezvous with the CEV and transfers its crew. Then the CEV heads back to earth for an Apollo-style reentry and parachute landing (at sea or on land). One version of the plan calls for transfer vehicles to switch constantly between earth and lunar orbit, carrying CEVs and lander modules back and forth. Similar hardware will deposit components for a moon base ahead of the astronauts' arrival. It should not be difficult to set up a permanent base. NASA has been studying how to do this for many years.

“There’s no doubt about it, space is a fabulous frontier, and we are going to solve some of its secrets and bring back many of its riches in my lifetime. I wouldn’t miss that for anything.”
Mercury astronaut Scott Carpenter, 1962

In the movie *2001: A Space Odyssey*, a huge manned spacecraft explores the Jupiter system. In truth, radiation hazards and financial realities make such a mission unlikely.

On the moon

> But is the moon the right place to go? Do we need a human presence there? According to the CEV's supporters, we must gain experience of living on the moon before we even consider doing the same on Mars. And anyway, the moon is a worthwhile target in its own right. One priority will be to build a super-sensitive astronomical telescope inside a crater, shielded by the moon's bulk from the electromagnetic "noise" of the Sun and earth. Mineral wealth will be another target. Significant discoveries may spur private industry to follow in NASA's dusty lunar footsteps.

During 2005, data gathered more than a decade ago by the uncrewed Clementine space probe was reanalyzed to produce a map of the moon, revealing the percentage of time that the surface is illuminated by the Sun during the lunar day (which lasts two terrestrial weeks). This may sound like just a technical scientific enquiry, but the results could determine the exact location of our first permanent moon base.

In full daylight the surface can reach blistering temperatures of 100°C, while at night it plunges to -180°C. This environment would place great stresses on a moon base. The long nights (again, lasting two terrestrial weeks) would also play havoc with solar power supplies. But there is at least one place that stays almost permanently in daylight and benefits from a relatively benign temperature all year round of -50°C: perfect for a base designed for long-term habitation. The magic spot is on the rim of Peary crater, close to the moon's north pole.

Another line of enquiry examined an opposing phenomenon: regions that are shrouded in permanent shadow. In 1998 a neutron spectrometer aboard the Lunar Prospector spacecraft scanned the topsoil for hydrogen-rich compounds. High readings in the flanks of some deep craters hinted at (although did not prove) the presence of water ice, shielded from evaporation because sunlight never strikes those sites. NASA suggests some of this could be mined for drinking water, and the rest broken down by solar-powered electrolysis into oxygen for life support and hydrogen for fuel. According to Cornell University astronomer Donald Campbell: "This profoundly affects the economics and viability of a lunar base. Water is the stuff of life."

Campbell reached this optimistic conclusion in April 2004, when the US Congress convened a hearing entitled Lunar Science and Resources. Others who gave evidence were not so convinced. John Lewis, a space chemistry expert from the University of Arizona, warned of "the immense difficulty of mining steel-hard permafrost in permanent darkness at the bottom of steep craters". It might be easier and less expensive to haul life support supplies from earth.

At any rate the ice, if it exists, is just a potential fuel supply. It is not necessarily a reason for building a lunar base in the first place. So, what about digging for valuable minerals? Supporters of NASA's plans claim that the moon's natural resources can be profitably exploited, but Texan astronomer Daniel Lester told the Lunar Science and Resources hearing that "few native materials may be of sufficient abundance, accessibility and value to merit their extraction and export. The costs would be prohibitive."

Some studies do suggest that lunar helium-3, deposited in the topsoil by charged particles in the solar wind, could be worth the effort of refining it. Helium-3 can be used in nuclear fusion to yield vast amounts of clean energy. Just 20 tons could power all of North America for a year. The trouble is, fusion reactors have not yet been proven to work, and no one is certain that helium-3 could be collected without shifting thousands of tons of lunar soil.

A moon base may be better for pure research rather than commerce. Since the dawn of the Space Age half a century ago, many astronomers have dreamed of building optical telescopes and radio dishes inside deep craters where they are shielded from solar light pollution and unwanted interference from earth's countless radio sources. Again, not all astronomers see the benefits. Daniel Lester suggested that the moon may well be fascinating for mineralogists but "as a telescope platform it offers mainly dirt and gravity", both of which degrade performance in comparison with free-flying instruments like the Hubble Space Telescope. As for the much-vaunted "radio-quiet" lunar sites for radio observations, Lester is skeptical. "The scientific need for them has never been entirely persuasive."

As we turn our longer-term attention towards Mars, NASA asserts that the moon is the best rehearsal ground and staging post. Unfortunately, most Mars champions don't see any need for stopping off along the way. "The moon is not 'between' earth and Mars," Lewis told the hearing. "It's a different destination."

Experts obviously disagree on what purpose a moon base might serve. But most space insiders believe one will be constructed soon, perhaps as early as 2020. Engineers are clearer about the challenges of making it than scientists are about the reasons for doing so. Europe's space agency ESA, for instance, has created the Moon-Mars Habitat Design Workshop, where young students from a dozen countries investigate a broad range of issues, from life support and psychological problems to the challenges of loading components into a moon-bound spacecraft.

NASA, meanwhile, has abandoned its antique dreams of giant lunar colonies built like lavish shopping malls. Today, lightweight inflatable structures are in fashion, accommodating just a handful of explorers. Pushed out of the hatchway of a lunar landing vehicle and then inflated, these could provide instant living space. Lunar dust shoveled on top would provide radiation shielding and micrometeorite protection, as well as helping to stabilize the interior temperature.

The lunar tidal pull is a strong one. In keeping with so many ambitious adventures that have been driven by our instinct to explore, we may only discover why we are on the moon once we have arrived.

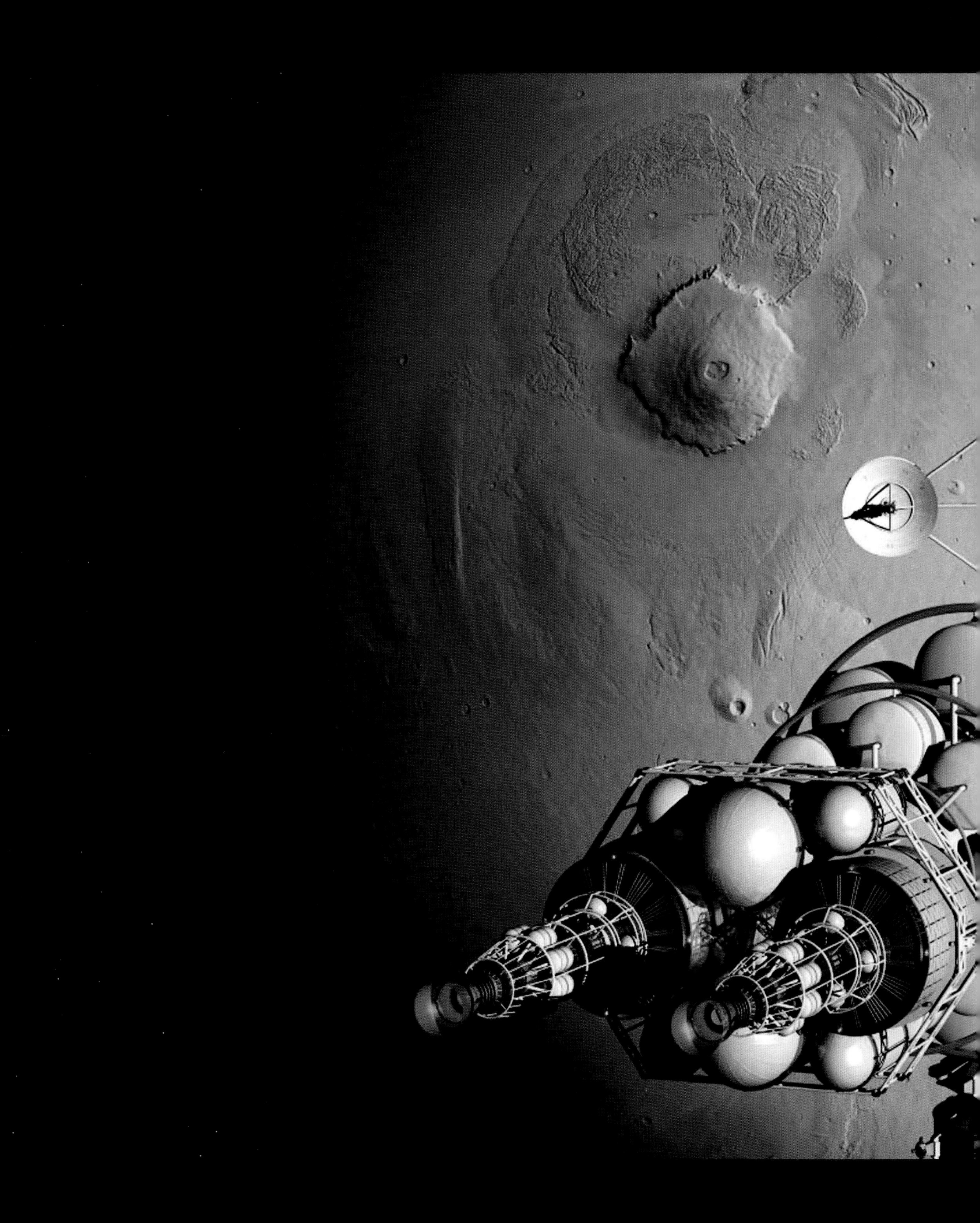

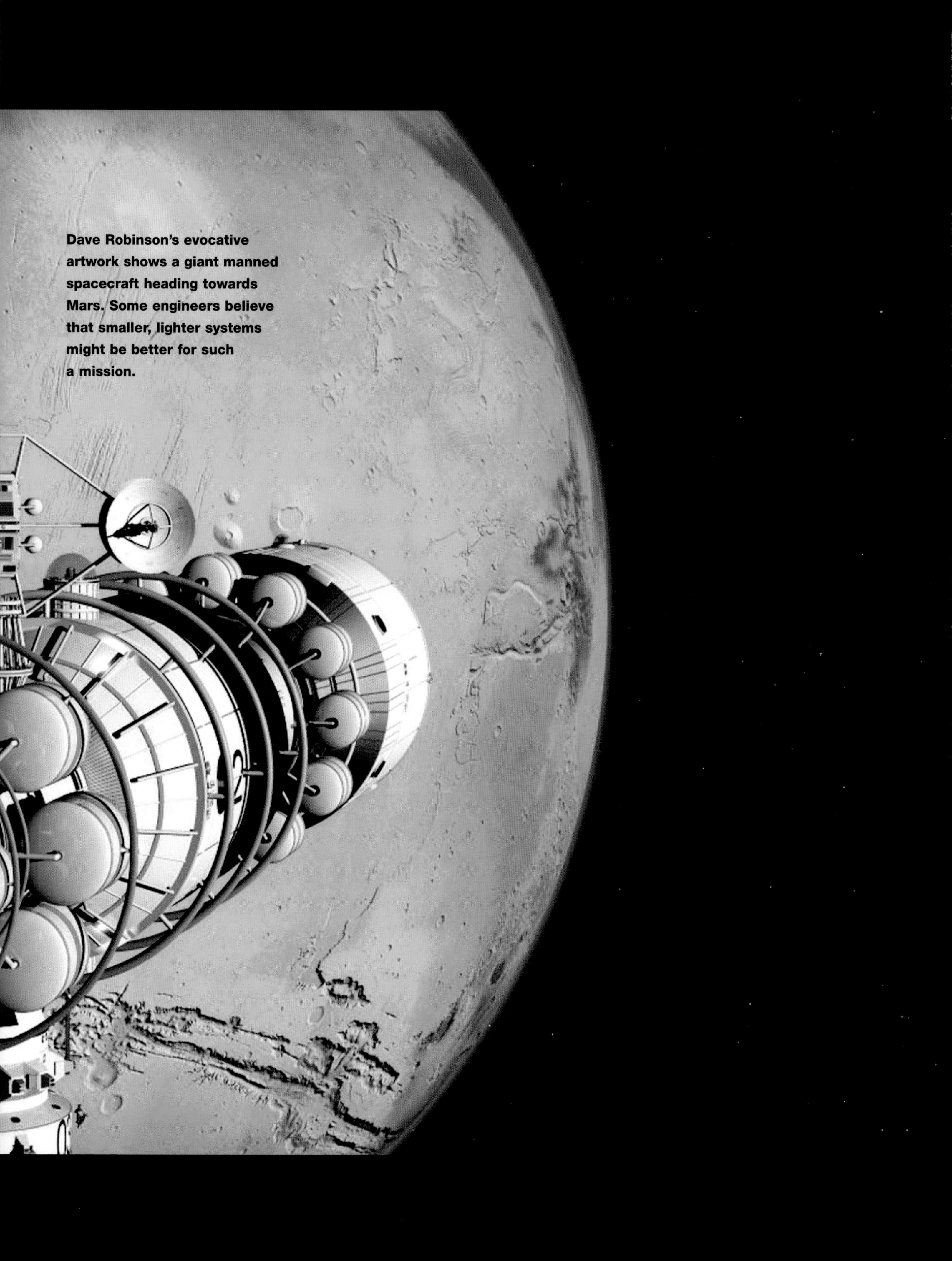

Dave Robinson's evocative artwork shows a giant manned spacecraft heading towards Mars. Some engineers believe that smaller, lighter systems might be better for such a mission.

The Lure of Mars

> Forty years ago, a senior Apollo manager called Joe Shea was listening to a presentation from one of NASA's contractors. They were explaining the various tests that had been conducted on a range of materials to clad the outside of the spacecraft and protect it against the fearsome heat of reentry. "They came to me and said, 'We've got a problem. We need a new heat shield.' They didn't know how to do it. I asked, them, 'How much extra money?' And they said, 'At least $60 million.' So I thought about it, and I said, 'Well, why is this the requirement?'" It turned out that the heat shield, although perfect for the searing temperatures of reentry, was not suited to the freezing temperatures encountered in space. "If one side is always facing the Sun and the other side always faces cold space, you have these thermal extremes." The shadowed side of the spacecraft would be so cold, and the sunlit side so warm that the heat shield would shatter like a china plate taken out of the fridge and then dipped in hot water.

Some $60 million of unexpected new expense. Many weeks of tests delaying the schedule. Possible unknowns added to the already daunting complexities of Apollo. In a flash, Shea saw the answer. "I came up with what I called the 'rotisserie mode'. I said, 'Let's just spin the spacecraft very slowly, once a minute or less.'" In this way, warm sunlight kept all areas of the craft at an even temperature. The entire problem was solved in as uncomplicated a way as possible. Today, some engineers are wondering if NASA can apply that kind of thinking to simplify a human mission to Mars.

Until recently, most mission plans have been so complicated and costly they have failed to win wide political support. On July 20, 1989, President George Bush Snr. celebrated Apollo 11's 20th anniversary. With Neil Armstrong, Buzz Aldrin and Mike Collins standing at his side, he said: "For the1990s, we have the space station. For the new century ahead, we should go back to the moon." Then, in words that were music to NASA's ears, he talked of "a journey into tomorrow, a journey to another planet, a manned mission to Mars". Short of setting a precise target date, he seemed to have given the green light.

A huge NASA team set to work on an infamous document nicknamed the "90-Day Report", so-called because this was how long it took to deliver a plan almost guaranteed to make the President wish he had never mentioned Mars. An expanded version of the space station was to service a thousand-ton interplanetary craft assembled in earth orbit. The Mars return trip would take eighteen months, with only a few weeks spent on the surface—just enough time to plant a flag and snap some photos before heading for home. NASA priced the project at just under $200 billion over two decades, but many financial analysts expected the figure to be more in the region of $500 billion.

At the Martin Marietta company, engineer Robert Zubrin was appalled by the costs NASA had in mind (as was the U.S. Congress, who gave no support to the idea). It seemed to him that building a giant spaceship had become more important than the Mars

mission itself. "I fired off a memo saying it wasn't enough simply to reach our destination. We had to do something useful when we got there. I thought the plan was totally wrong and too expensive, and many people at NASA were upset when I spoke out of turn." Industry didn't appreciate what Zubrin had said either. "Aerospace companies usually tell NASA exactly what they want to hear, because that's the way to make a sale. I was proposing to do the opposite and tell the truth, whether NASA liked it or not. Theirs was the worst and most inefficient way to get to Mars."

Al Schallenmuller, Marietta's chief of civilian space systems, fondly remembered working as an engineer on NASA's robotic Viking lander missions of 1976, and he was keen to see a manned Mars mission if it was at all possible. He allowed Zubrin and a dozen colleagues some time to rewrite the company's official sales pitch to NASA. By February 1990, the team had reduced the space station's role in the Mars mission, cut the weight of the outbound ship in half and slashed the costs.

But Zubrin was not yet satisfied. Time spent on the Martian surface was still only four weeks out of an eighteen-month round-trip. The earth orbital construction of the Mars ship annoyed him because it cost money and wasted fuel; and the ship was still much heavier than he had hoped. He came up with "Mars Direct", a plan that cut size, weight and costs to the bone, required no time in earth orbit and eliminated the International Space Station's role as a construction base and stopping-off point.

Intelligent roving robots will extend the reach of human Mars explorers.

"Most Mars plans call for a huge mothership to circle the planet and send down small landing teams, which then come up, rendezvous with the ship and fly home. I call it the 'Battlestar Galactica' approach. Why have the mothership at all? In Mars Direct, you fly hardware directly from earth to Mars, and then a small earth Return Vehicle (ERV) fires off the surface and heads back home." This smaller style of ship would mean confining astronauts to relatively cramped cabins for the six-month outward and eight-month return trips, "but we know from the Mir and space station experience that people can tolerate that if they're sufficiently motivated. We don't have to build giant space cruisers to go to Mars."

A huge amount of fuel (and therefore spacecraft size and weight) is saved by launching only when earth and Mars swing close to each other in their orbits and are both on the same side of the Sun at once. This happens roughly every two years, so Mars Direct employs a rolling schedule of missions to coincide with these close planetary approaches. The downside to this "low-energy" scheme is that each mission takes well over two years from start to finish, because the crew has to wait eighteen months on Mars until the planet swings close by the earth again before they can set off on the return trip. On the face of it this lengthens their exposure to cosmic radiation hazards. But while Mars Direct involves a longer stay on the planet, astronauts actually spend a little less time in deep space, where the radiation hazard is greatest.

Supporters of Mars Direct also point out that space radiation does not kill people. What it actually does is increase the risk of radiation-related cancers in later life. So far, the overall death rate of astronauts shows that (accidents aside) their longevity is no different from normal. "We've spent more than forty years sending astronauts to the moon or putting them in space

Inflatable modules could allow large-volume Mars bases to be stored compactly during the outward flight from earth.

"The question of taking on Mars as an interplanetary goal is not simply one of aerospace accomplishment, but one of reaffirming the pioneering character of our society."
Robert Zubrin, *The Case for Mars*, 1996.

stations for very long periods," Zubrin argues. "There is nothing much more we can learn about the radiation risks that we don't know already."

Weightlessness is another problem. We know that astronauts on extended space missions lose body mass, muscle tone and bone strength. Here again, lateral thinking might ease these difficulties on a Mars mission. A long tether stretching between two modules can create a basic version of the gently rotating artificial-gravity space stations so favored by rocket visionaries in the 1950s; only instead of a spinning wheel, the ship would look more like a tumbling stick. Astronauts would arrive at Mars in pretty good shape, yet without having to travel in some vast and complicated starship. For the return voyage only one module would be used, so the tether and its gift of artificial gravity would be lost. If as a result astronauts were a little less than super-strong on their return, it would not matter. There would be reception teams and plenty of support on standby. Mars explorers would land on earth as easily as astronauts coming home after long missions aboard space stations.

There is always a chance that a crew might get stranded. What if their Mars lander crashed, or failed to take off at the end of their stay? Again, this is a simple problem to fix. Uncrewed earth Return Vehicles are launched to Mars under computer control. The modules then touch down and report back their status. Only when they are sitting safe on the surface, ready to lift off for home, are humans sent to join them. In addition, these modules could land with their fuel tanks half empty, thus saving even more weight and reducing the cost of launching them from earth. The key to this saving is that Mars's atmosphere consists mainly of carbon dioxide. An "in-situ resource utilization" (ISRU) plant could pump this through a nickel catalyst, adding a trace of hydrogen into the mixing chamber. The catalyst splits the carbon dioxide, liberates the oxygen and combines it with the hydrogen to make water. The freed carbon reacts with spare hydrogen to create methane—rocket fuel for the return trip. Six tons of hydrogen carried to Mars could be converted into 108 tons of methane and oxygen. No new technology has to be invented to make ISRU work, and the technique has already been proven. Indeed, future robotic Mars probes will carry experimental ISRUs.

One by one, Zubrin and his army of supporters in the highly organized Mars Society lobbying group are trying to knock down the anxieties and cost implications that, so far, have prevented astronauts from flying to Mars. NASA is now listening.

But old space agency hands recall that Joe Shea, the brilliant, passionately dedicated man who brought his logic and ingenuity to project Apollo, was also one of the many NASA people who failed to spot "obvious" fire hazards in the Apollo 1 capsule. There will be no easy answers in our quest to go to Mars, and there are bound to be some unpleasant surprises as well as sweet victories. There is no guaranteed formula for success, only a varying degree of risk to be played off against time, cost and political will.

A pressurized roving vehicle serves as a mobile self-contained Mars base in this visualization of a NASA proposal.

Once again, we return to the issue of NASA's ability to communicate the real nature of space exploration to the public. An element of danger must be acknowledged, even welcomed, so long as all the hardware is designed in good faith. This fundamental honesty will not only heighten the human drama of a Mars flight, it may also bring the costs of a mission down to manageable levels while allowing NASA's spacecraft engineers greater freedom of expression.

The Mars Society has attracted corporate sponsorship to build trial Mars habitats in some of earth's hottest, driest and coldest locations. Teams are already experimenting with the disciplines relevant to a real mission. In January 2003, President George Bush Jr. announced that Mars should be the logical destination for a new "Vision for Space Exploration". His advisors were determined not to repeat the mistakes of the 90-Day Report in 1989. The first human footprint on Mars may—perhaps—be one small step closer.

But it will still be incredibly expensive and there are many scientists who argue that the slow and costly astronaut program is actually getting in the way of true space exploration. It might be better to explore the solar system with expendable machines whose safety we don't have to worry about. Judging from the results so far, those scientists may be right. Robot probes have investigated incredible alien worlds in the farthest reaches of space. And today we can all share those wonders at the click of a computer mouse.

ATRV - JR
REAL WORLD INTERFACE

08 ROBOT AMBASSADORS

"I remember being transfixed by the first lander image to show the horizon of Mars. This was not an alien world, I thought. I knew places like it in Colorado and Arizona and Nevada. There were rocks and sand drifts and a distant eminence, as natural and unselfconscious as any landscape on earth. Mars was a place."
Carl Sagan, *Cosmos*, 1980

ROBOT AMBASSADORS

> When the Italian astronomer Giovanni Schiaparelli drew the first map of Mars in 1877, he had nothing more to guide him than the blurred images captured by his simple telescope. He recorded, as best he could, several large dark plains loosely connected by much narrower features, which he labeled canali. In Italian, this simply means a groove or channel. An enduring myth was created when Schiaparelli's work was translated for English-reading astronomers. The word canali was interpreted as "canals"—artificial water-bearing structures created by an intelligent civilization.

In 1894 a brilliant but misguided American astronomer, Percival Lowell, made observations of Mars from a high-altitude observatory in Flagstaff, Arizona. He was sure he could see canals crisscrossing the entire Martian surface. In 1896 he published his findings. "On Mars we see the products of an intelligence. There is a network of irrigation... Certainly we see hints of beings in advance of us."

What might Lowell have "seen" if he had never encountered those flawed translations of Schiaparelli's findings? We will never know. Lowell's work was derided by other astronomers, but his misapprehensions turned out to be much more entertaining than the reality. His proposal that there existed on Mars an ancient water-starved civilization that had girded an entire world with canals and pumping stations was too good to waste. Newspapers and popular writers took up Lowell's ideas with enthusiasm, not least H.G. Wells in his classic novel *The War of the Worlds* (1896), the tale of an invasion of earth undertaken by intelligent but merciless Martians. The novel's opening lines retain all their power today:

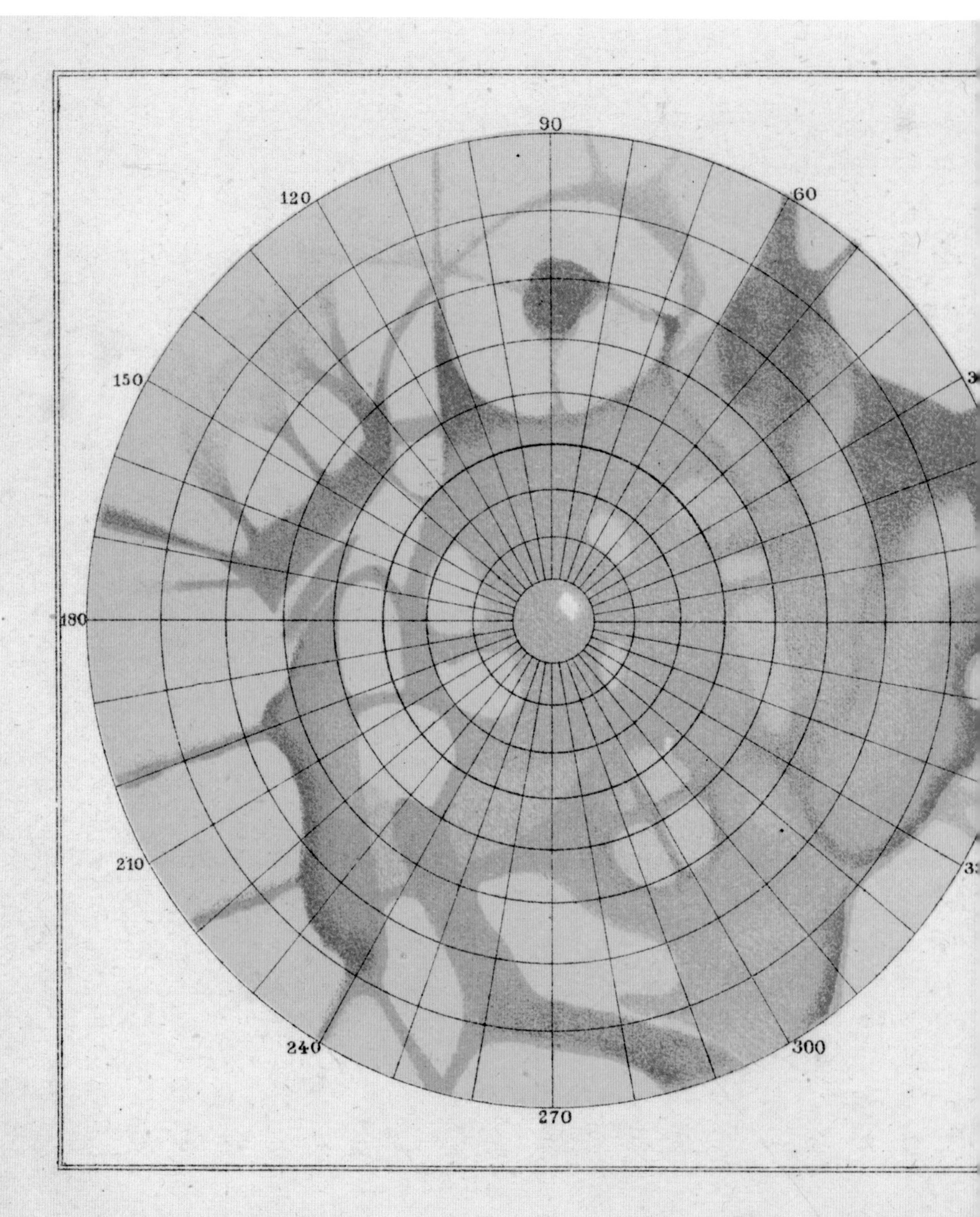

Die Hemisphä

Nach J.

Südliche Hemisphäre.

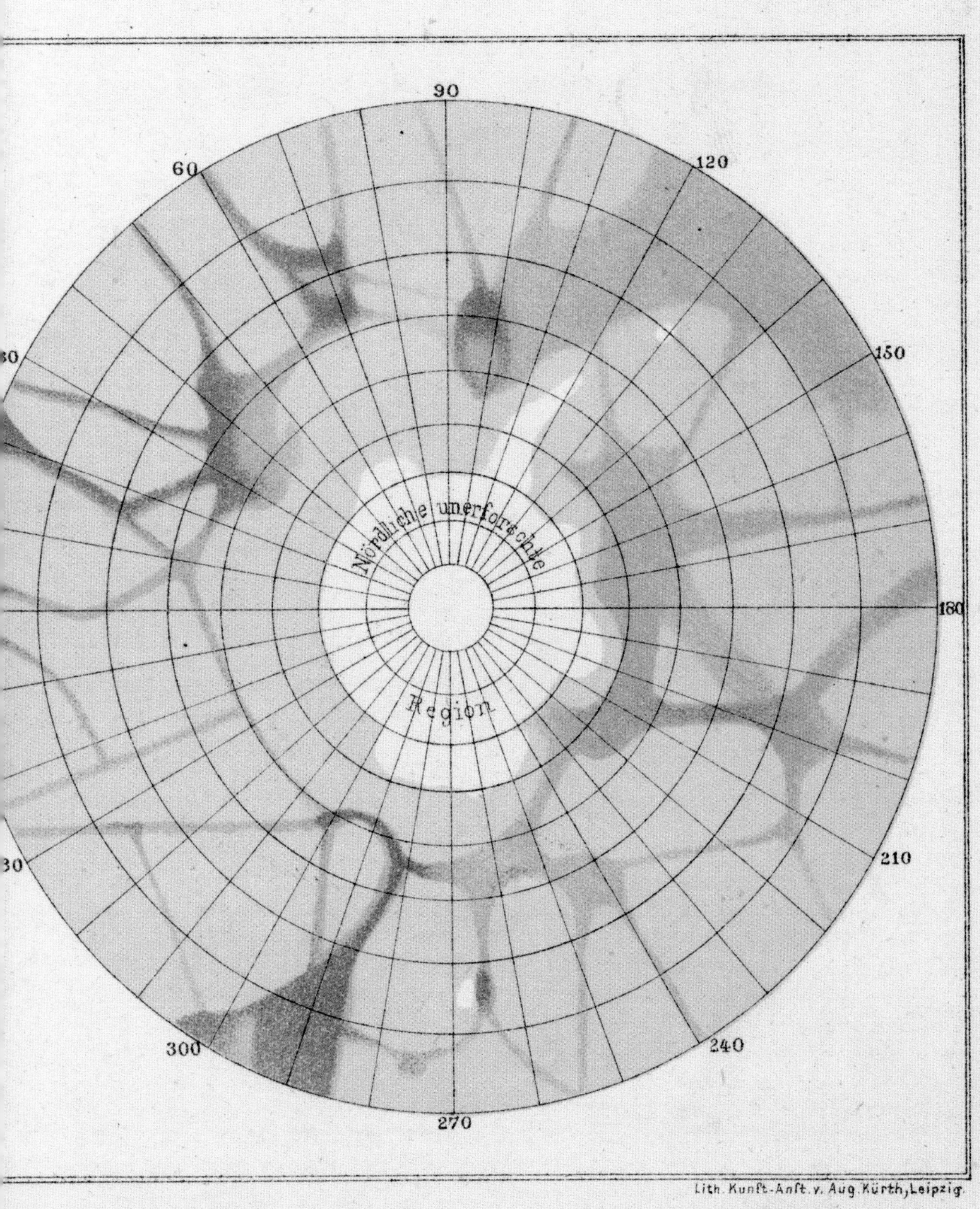

des Mars 1879.

iaparelli.

Nördliche Hemisphäre.

This 19th century "map" of Mars was among the best that astronomers could deliver at the time.

"There's nowhere we cannot go. And in that belief, we set out for other worlds, all brimming with confidence. And what are we going to do with them? Rule them, or be ruled by them. This is the only idea in our pathetic minds. What a useless waste!"
Stanislaw Lem, *Solaris*, 1961

"No one would have believed in the last years of the nineteenth century that this world was being watched keenly and closely by intelligences greater than man's and yet as mortal as his own...Across the gulf of space, minds that are to our minds as ours are to the beasts that perish, intellects vast and cool and unsympathetic, regarded this earth with envious eyes, and slowly and surely drew their plans against us."

By the end of the nineteenth century, astronomers had acquired a new tool, the spectroscope. Analysis of sunlight reflected from the Martian surface, or glancing obliquely through its atmosphere, provided some sobering data. The air was whisper-thin, incredibly cold, and comprised mainly of carbon dioxide. Neither Lowell nor the fiction writers allowed these observations to interfere with their grand visions of an intelligent civilization. The dominant twentieth-century image of Mars was promoted in cheap story magazines. Edgar Rice Burroughs, creator of the jungle hero Tarzan, wrote a series of fantasies set on Mars (called Barsoom by its inhabitants). Just as Wells had been, Burroughs was greatly influenced by Lowell's ideas. Through his cheap and cheerful stories, the reading public soon developed a taste for life on the Red Planet. Barsoom's canals were the least of its attractions. It also boasted beautiful princesses.

Mars hit the headlines in 1938 when Howard Koch and Orson Welles turned *The War of the Worlds* into a radio play that included a realistic news broadcast. Many Americans thought the bulletin was genuine and that Martian invaders had landed in New Jersey. In 2005 the film maker Steven Spielberg made a spectacular and unsettling version of *The War of the Worlds* aimed at a post-9/11 audience.

Science fiction has its uses on earth, but real-life discoveries are not always so flexible. During the 1950s most scientists reconciled themselves to the fact that Mars was not a likely abode for intelligent forms of life. However, they still expected something. As the space planners in America and Russia developed their new and potent rocket technologies, Mars became a favored target. A human mission seemed possible, in theory at least, for the next generation. In *Das Marsprojekt* (1952), Wernher von Braun outlined how 1950s technology might achieve a mission. Ten ships, 4,000 tons apiece, would make the trip.

Arthur C. Clarke wrote a fine science fiction novel, *The Sands of Mars* (1951), depicting a self-sufficient colony of earthlings breaking away from their mother planet: a theme that continues to haunt scientific and literary

Top, left and right, the two small images of Mars, taken by a Mariner spacecraft in 1964, were an incredible achievement at the time. The larger image, below, of the Ophir Chasma region, taken by the ESA spacecraft Mars Express, shows what we are capable of today.

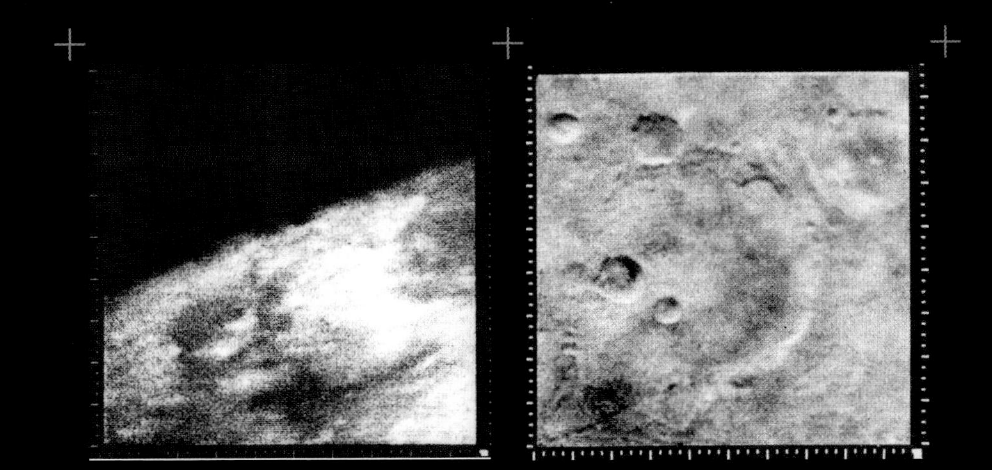

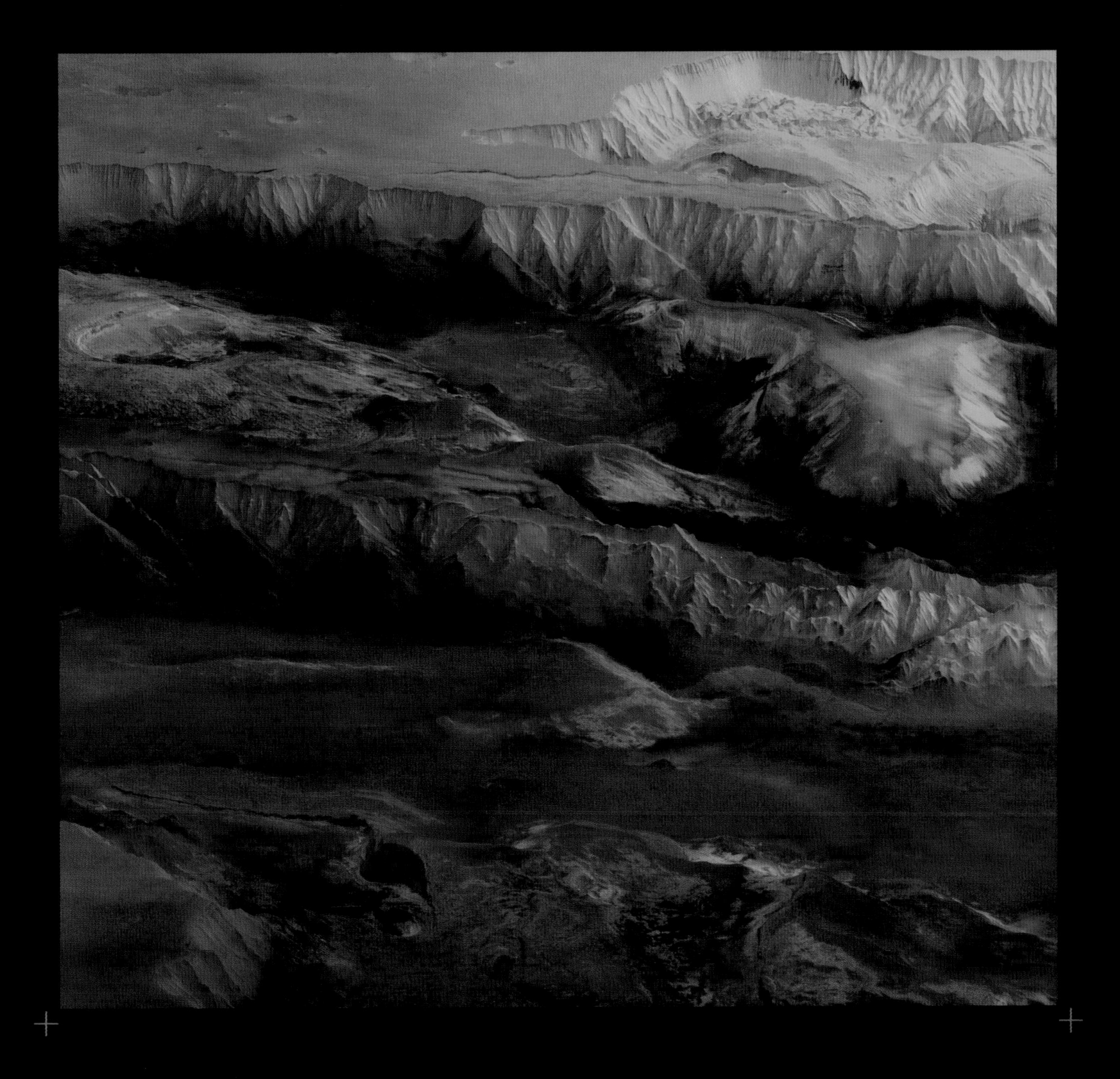

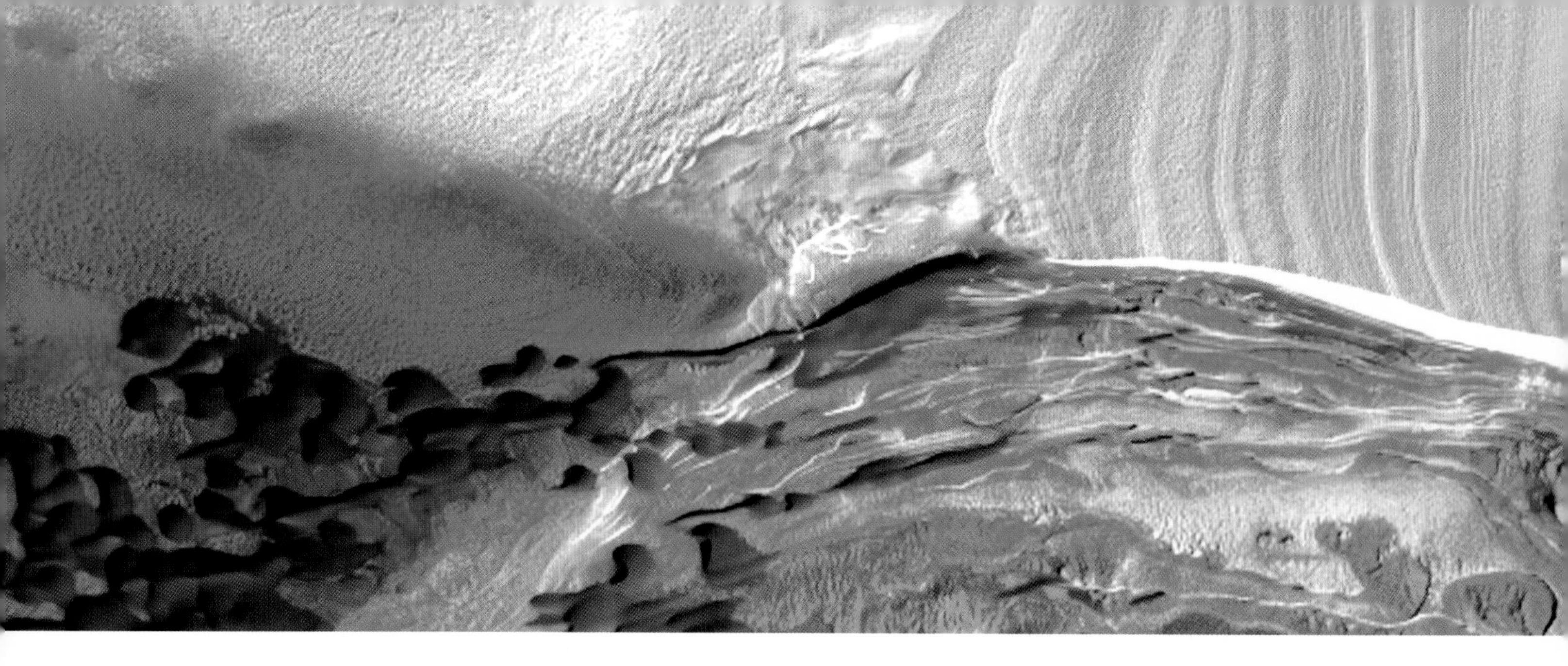

imaginations. Ray Bradbury used the planet as the setting for one of the most famous science fiction works of the twentieth century, *The Martian Chronicles* (1950), in which Mars was inhabited by the ghostly remnants of an alien culture. Bradbury also depicted the carelessness of humans when they enter a new and unknown environment. Robert Heinlein promoted his hopes for the betterment of mankind by using Martian intelligence as a model in his novel *Stranger in a Strange Land* (1961). Its influence fed into the 1960s counterculture.
So when did our current perception of Mars first take hold of our imaginations: the dusty, desolate, boulder-strewn world we know today? By the 1960s, Lowell's canals had been firmly ruled out by the first robotic space surveys. However, in a report entitled "Conquering the Sun's Empire" prepared by NASA in 1963, the authors stated: "We can feel reasonably confident that primitive life exists." A decade later, a pair of complex Viking robot landers sampled the Martian surface soils, using instruments designed specifically to detect life. The results were inconclusive and for a while it seemed as if the long-held dream of finding even the simplest creatures on the Red Planet had died in the cold, thin, ultraviolet-soaked atmosphere and dry soils encountered by those Vikings.

For nearly two decades after the Viking mission, the world seemed to lose interest in Mars. During the late 1960s, trajectory experts were starting to look further into space at the gas giants, such as Jupiter and Saturn. They calculated that earth and the outer planets would all pass around the same side of the Sun between 1977 and 1978. This happens only once every 176 years, so it was an opportunity not to be missed. NASA planners wanted to create a mission that would use the gravity fields of Jupiter, Saturn, Uranus and Neptune to alter the course of a robot probe as it swung by, so that each world would divert the ship to the next. The plan was nicknamed the "Grand Tour".

With the probe dragged towards a given planet by gravity, it would pick up velocity and then hurtle around the planet's far side. By mapping out the initial approach very carefully, the flight dynamics experts could ensure that the angle of deflection in each case would enable the ship to continue towards the next planet with little need for rocket engines. This meant that the flight time of a conventional rocket from earth to Neptune could be reduced from thirty years to twelve, a length of time perhaps within the operational life of a hardily constructed vehicle.

However, the budget analysts instructed NASA that there were funds to explore Jupiter and Saturn only. The planned Voyager probe, costing half as much as Viking, could not be equipped with instruments for any extended mission. But the hardware designers did their best to construct the twin Voyagers to be as sturdy as possible. Just in case.

The ships were launched a month apart, in August and September 1977. Two years later they swung past Jupiter and returned stunning images of a world in

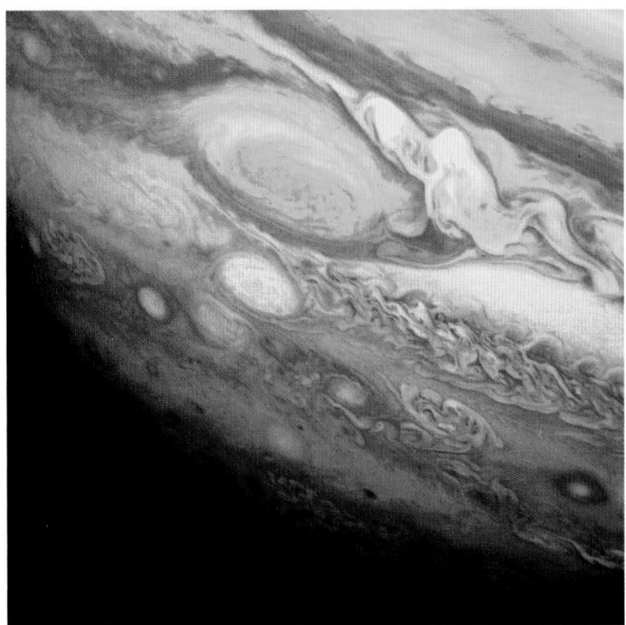

turmoil, with banded clouds, a mass of strange colors, and a permanent red-tinged hurricane that could swallow several earths.

In the light of these staggering successes, NASA was authorized to extend the mission. Both Voyagers went on to Saturn, where they encountered a ring system more complicated than anyone had expected, as finely grooved as an old long-play record. Some of the finer rings twist around each other like braided ropes as a result of gravitational interactions that we have not even begun to understand properly. Voyager I was diverted from its planned course so that it could study some of Saturn's moons, while Voyager II continued outwards to Uranus (1986) and Neptune (1989).

After their main planetary encounters, the Voyagers continued to transmit data about the interplanetary environment. In 1996 the decision was taken to shut down the fading radio link with the probes. They had escaped the Sun's grip and left the solar system. As Voyager I slipped the bonds of Saturn's gravity and drifted into deepest space, its last picture showed the earth and six other planets as tiny sparks in the darkness, shepherded by a single bright but distant star, the Sun. The astronomer and science popularizer Carl Sagan had lobbied vigorously for this, even though NASA's camera team was worried that their delicate instruments might be damaged if they looked back towards the Sun's glare. The image showed our home planet in its true perspective, as a "pale blue dot" in a vast emptiness of dark. It was only by virtue of a handy arrow on the press release that anyone could distinguish our particular dot of light from countless other anonymous little sparks drifting in the cosmic blackness. Once again, a space mission had brought a new perspective on this world even as it revealed alien ones.

In 1993 Mars was back in the headlines, although for all the wrong reasons. A new craft called Global Explorer was supposed to map the planet's surface in precise detail. On August 21, the probe was getting ready to fire its retro rocket in order to slow down and allow the Martian gravity field to capture it. Then, just as the fuel tanks were being pressurized, NASA lost its signal. The $2 billion machine probably exploded when its fuel tanks failed.

Then, in July 1996, a team of NASA scientists announced that they had discovered possible traces of fossilized Martian life... on earth.

On December 27, 1984, a small group of explorers had been driving across an Antarctic ice field in a little powered sled, hunting for meteorites (which can easily

Opposite, the Martian north pole, showing layers of ice, complex erosion patterns, and dark deposits from sand storms. Above, Jupiter's most famous and long-lived atmospheric storm, the Great Red Spot, captured by the Voyager 1 space probe in 1975.

Mars rover Spirit photographs its own shadow. In the next couple of decades, super-intelligent planetary probes might actually become aware of their surroundings.

be identified on the otherwise rock-free ice sheets). The sled stopped and a parka-clad young woman stepped out. Something had caught her eye: a small dark lump the size of a potato, lying half-buried in the ice. "Hey, this looks like a good one!" she shouted. More than a decade later, Roberta Score, a member of a National Science Foundation meteorite hunting team, would find herself in front of dozens of cameras in Washington D.C., explaining what she had discovered in the remote Allen Hills ice fields of Antarctica. Why had she stopped the sled and instantly noticed that rock? "The colors looked different. The rock looked very green. It stood out in my mind that it was kind of weird."

Just how weird was not immediately apparent to the survey team. Sample ALH84001 was logged as a probable asteroid fragment. It was sealed into a special canister and shipped to NASA's Johnson Space Center in Houston, Texas. For nine years it sat in a protective sealed storage facility, surrounded by nitrogen gas to protect it against chemical corrosion. And then at last, in 1995, it was analyzed in more detail by scientists.

On Monday, August 5, 1996, the journal *Space News* published its usual reports on NASA's activities, rocket launch schedules and hardware. A snippet reported on a rumor circulating among the space community, headlined 'Meteorite Find Incites Speculation on Mars Life'. By Monday night the story had been picked up by the national media. The next day, NASA's chief administrator Daniel Goldin was forced, ahead of schedule, to put out a formal press release. "NASA has made a startling discovery that points to the possibility that a primitive form of microscopic life may have existed on Mars more than three billion years ago. The research is based on a sophisticated examination of an ancient meteorite that landed on earth 13,000 years ago."

Goldin was at pains to point out that all the findings were tentative and that more work needed to be done. "The evidence is exciting, even compelling, but not conclusive," he warned.

On Wednesday, August 8, President Clinton made a special TV announcement of what appeared to be a great achievement. "If this discovery is confirmed, it will surely be one of the most stunning insights into our world that science has ever uncovered."
Yet another decade has passed and still the debate about life on Mars—whether long vanished or still thriving—has not been resolved. The mineral clues in ALH84001 remain frustratingly open to different interpretations. Meanwhile, other targets in the solar system are attracting the attention of space biologists.

In August 1996 the Galileo space probe beamed back detailed images of Europa, Jupiter's fourth and largest moon. Its crust appeared cracked across the entire surface. Geologists believe that the surface melts and reconstitutes itself at intervals. Close-up views of the many fractures in the crust are certainly intriguing. There are parallel strips, with brighter strips running alongside darker ones. It seems as though some dark,

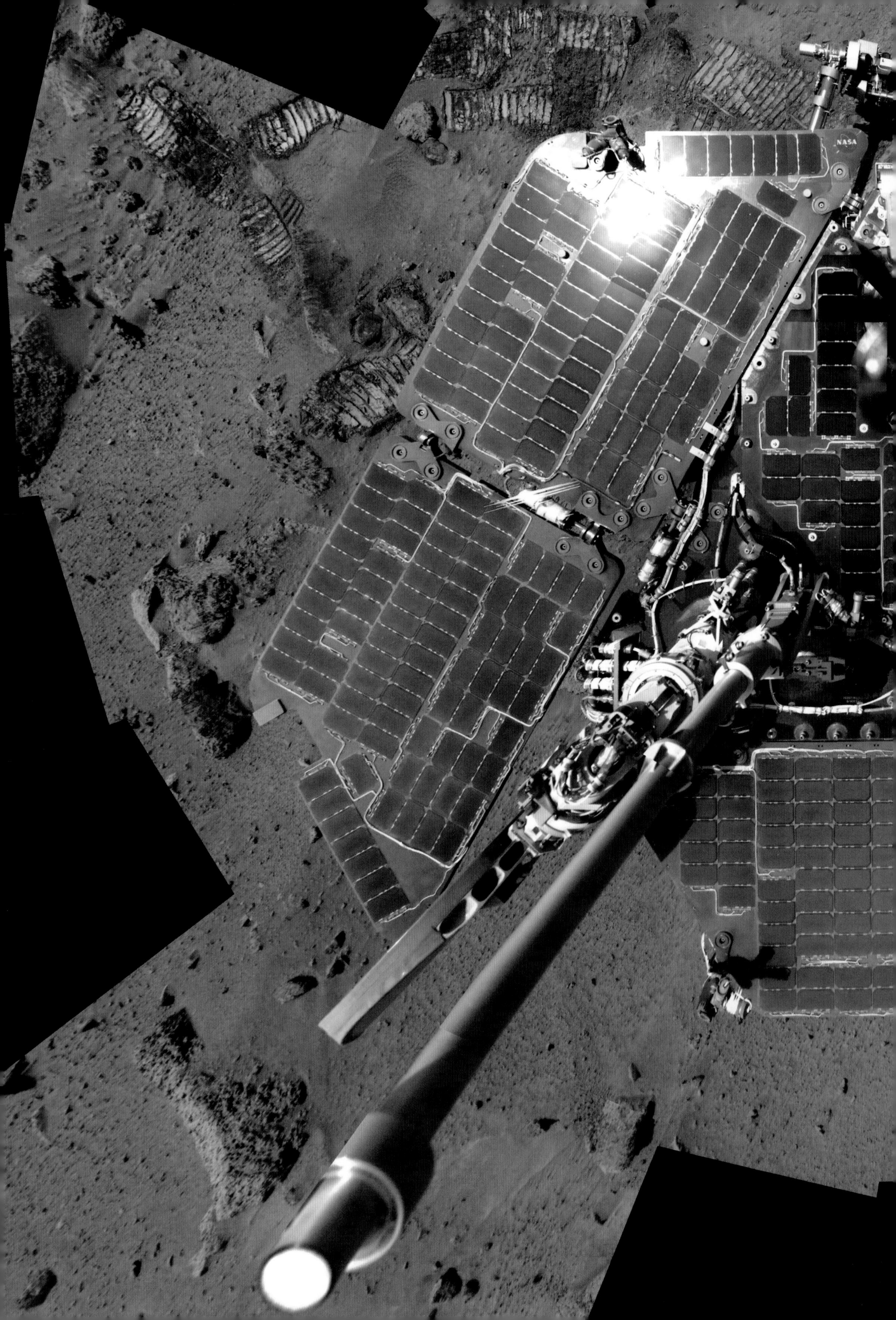
NASA

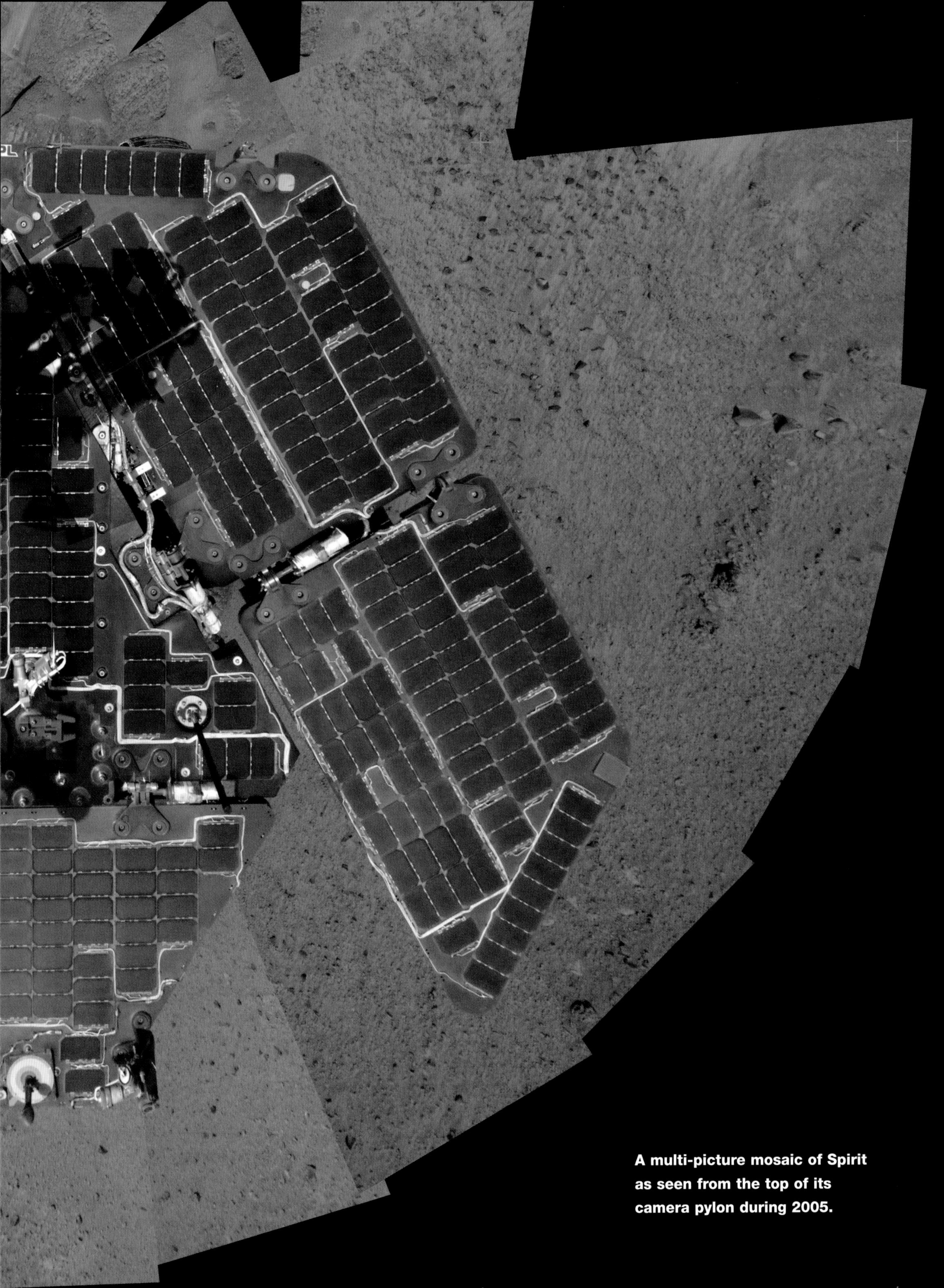

A multi-picture mosaic of Spirit as seen from the top of its camera pylon during 2005.

A landing stage and deflated airbag system from a successful Mars Exploration Rover (MER) mission, photographed by a rover as it prepares to leave the landing site and strike out for fresh horizons.

slushy material is oozing up through the cracks, leaving a stain. NASA believes that the crust is a thick layer of ice. Geologist Ronald Greeley describes the patterns as "just like ice floes on the earth's polar seas". He and many other scientists suggest that a huge ocean of fluid water, 30 miles deep, may exist just under Europa's brittle surface. Tidal forces from Jupiter prevent the water from freezing. Quite possibly, it is comparatively warm. When it pushes up between the cracks and freezes into place, it deposits grains of darker material. Intriguingly, Galileo's instruments detected possible traces of organic chemicals, the primitive building blocks required for life.

Fascinating chemistry may also await discovery on one of the moons of Saturn. On January 14, 2005, the European space probe Huygens touched down on Titan, a far stranger and more distant place than Mars. Huygens travelled through space for seven years, clinging to the flanks of NASA's Cassini craft (the largest interplanetary probe ever launched) as it swept across the solar system on a complicated two-billion mile trajectory. In that time, the combined craft looped once past the earth, twice past Venus and once around Jupiter in a complex series of "fly-by" maneuvers designed to boost its velocity and hurl it towards distant Saturn.

After its immense voyage, Huygens was designed to operate for just a few hours once it had been released from Cassini to free-fall into Titan's atmosphere. Those few hours were enough to provide a glimpse of an awe-inspiring alien landscape. The technical challenges of a touchdown on Titan were far greater than for depositing a craft onto the surface of Mars. This was a return to the earliest emotional heights of the Space Age, where once again, after many months and years of planning and flight time, an incredibly distant alien world could only be glimpsed for a brief few moments before the very best of our technological prowess was exhausted. The images returned from Titan disappointed some commentators more used to the full-color views of other, easier worlds. Color data would have required too much time and power to transmit from within Titan's dense and almost impenetrably opaque atmosphere. Yet, once again, their crude black-and-white imagery highlights the epic scale of the achievement.

As Huygens plunged through the sky of Titan, it confirmed what scientists have long suspected. The air is mostly nitrogen, but there's also a thick smog of methane clouds. When the cameras switched on at an altitude of 90 miles some real surprises were in store. ESA scientists at Darmstadt in Germany knew from the first frames that they had struck gold. As imaging team member Anthony Del Genio recalls: "We were lucky to come down smack over a region where you could see a wide variety of surface types in the same scene. That gave us a lot of perspective." The cameras showed that methane rain falls on the surface and liquid methane rivers scour Titan's landscape as surely as rain carves valleys on earth. (In Titan's frigid temperatures, water freezes hard as rock while methane condenses into liquid form.)

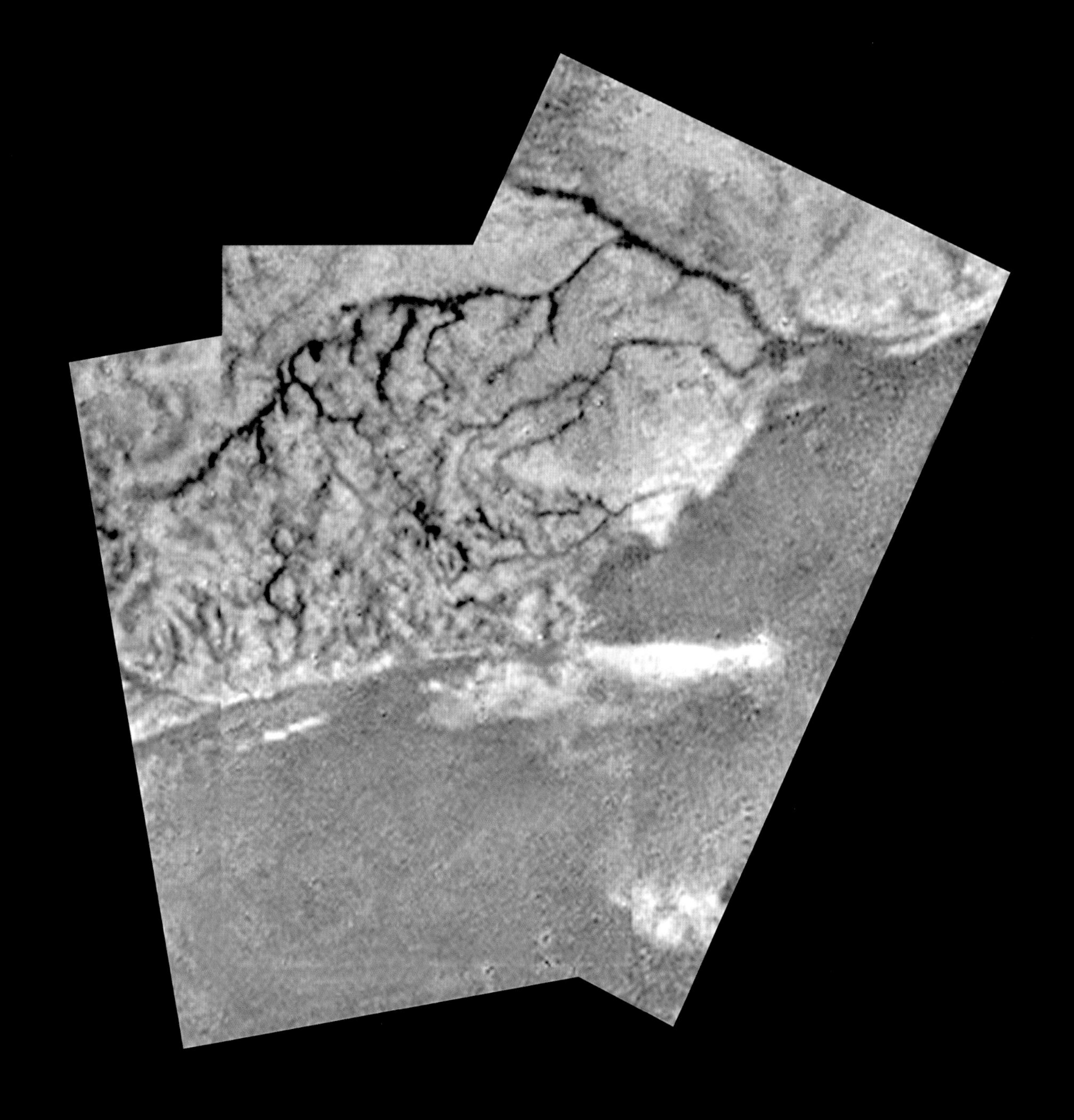

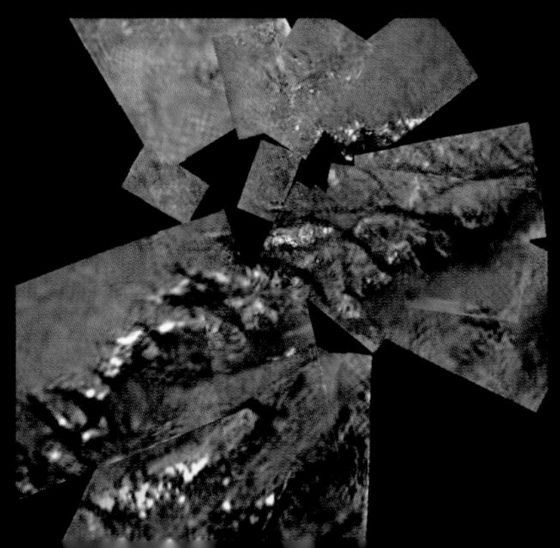

A mosaic of what may be a river of liquid methane on the surface of Titan. The river appears to flow down towards a huge lake

"In some ways, the machines we use to explore Mars are as intelligent as an entire university laboratory. In other ways they're no smarter than a grasshopper. You could describe a lot of people that way."

Planetary scientist Carl Sagan, 1993

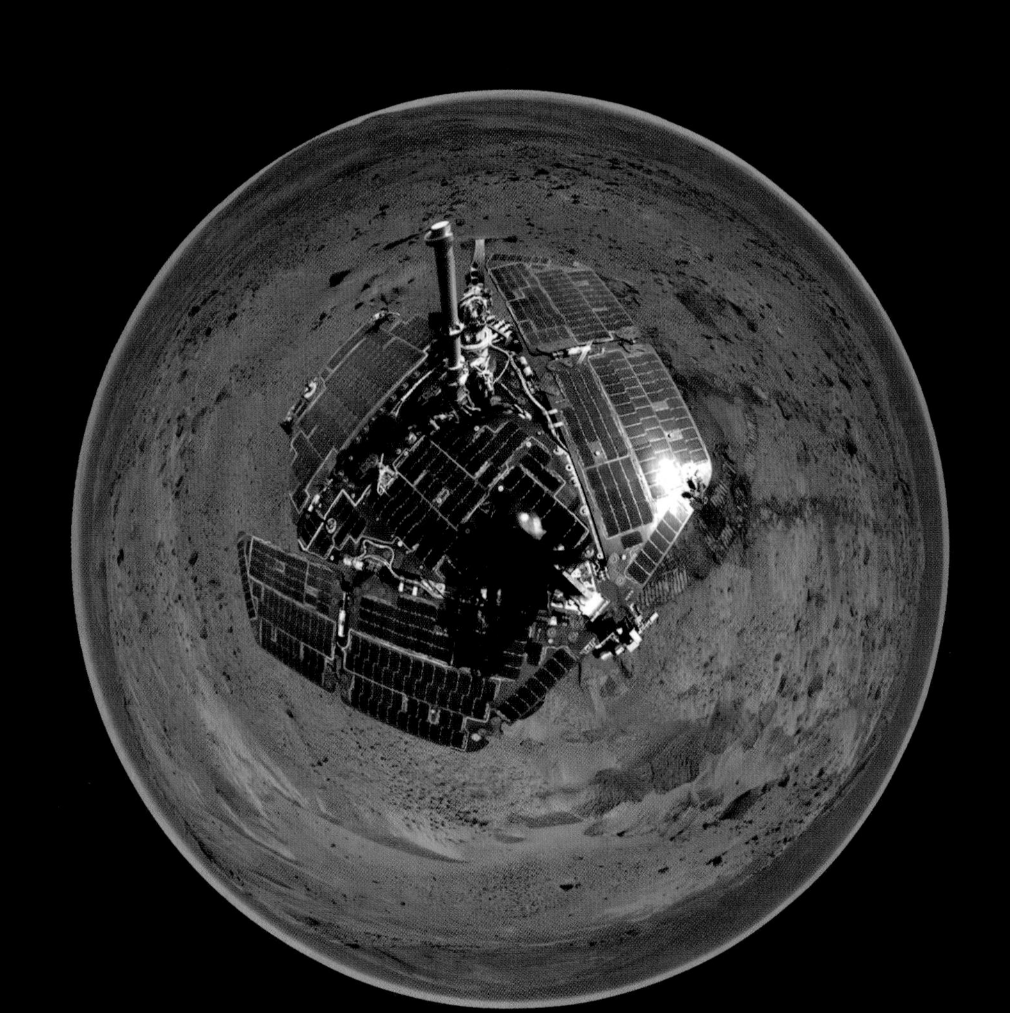

An artist's depiction of ESA's Huygens probe on the cold and inhospitable surface of Titan.

Right, a view of Titan's smog-shrouded surface.
Opposite, ESA's Ariane V launch vehicle is one of the most advanced rockets at work today. America and Russia are no longer alone in the leadership of space exploration.

But where do those rivers end up? From its vantage point further out in space, Cassini's infrared camera detected regions of light and dark texture beneath Titan's dense atmospheric haze. Closer inspection by Huygens showed that the brighter areas are rough terrain, probably consisting of water ice. Dark, low-lying zones are flat and have distinct "shorelines". So it looks as though the methane rivers feed vast lakes. They may not be a liquid you could swim in, but scientists are confident that they are lakes of some kind, perhaps filled with an icy slush rather than free-flowing liquid. Martin Tomasko, leader of the Huygens camera team, says: "We're seeing earthlike processes with very exotic materials. There's plenty of evidence for flowing fluids."

The color difference between the high ground, the river channels and lake beds also caused a stir. Although the sunlight hitting Titan is barely a hundredth as intense as on earth, it is still strong enough to trigger chemical reactions in the upper atmosphere, breaking down methane to create a haze of complex hydro-carbons. These molecules fall to the ground as a smoggy rain, leaving dark deposits on the icy ground. Titan appears to be rich in organic chemicals.

Humans are extremely keen to discover whether or not we are alone in the Universe, but there is a major problem to be solved before we can answer this question. Scientists cannot even reliably detect life from space on the one world where its signs should be glaringly obvious: the earth. If they can't find it here, they will have a tough job finding it anywhere else.

When the Galileo space probe swooped by earth in 1990, building up momentum for its long trip to Jupiter, all its instruments were pointed towards us for a unique experiment. Strong absorption of light at the red end of the visible spectrum, particularly over the continents, indicated the presence of chlorophyll, the molecule essential for photosynthesis and plant life. Spectral analysis of sunlight passing through the earth's atmosphere revealed a high oxygen content. Since oxygen is extremely reactive, a dead planet could not support free oxygen in its atmosphere for very long. It takes life (namely plants) to replenish it constantly. Galileo also spotted small quantities of methane in the atmosphere.

However, in August 2003 a perplexed NASA team reported their failure to sense life in the Atacama Desert of northern Chile. Admittedly, it is one of the world's driest deserts, but it is teeming with life nonetheless. Scientists on the ground were pestered by flies, even as they marveled at the variety of lichens growing on and under rocks, or watched vultures circling over their heads: a sure indication that plenty of other animals had to be around somewhere. But colleagues at the Ames Research Center in California, poring over photos and instrument data transmitted from the field scientists, did not detect anything they considered an unambiguous sign of life. These results were unsettling for scientists trying to develop robotic landers and instruments for future planetary exploration. What subtle data signatures, beamed to us on fragile radio links from an unimaginably remote alien planet, might prove its existence?

esa
esa
cnes
cnes

esa
esa

Mars@click.mouse

> There was a time, just four or five decades ago, when the planets in our solar system were almost unimaginably distant and strange to us. It seemed a great achievement if we could send simple space probes to obtain even a brief glimpse of these worlds. Entire missions were planned and launched whose purpose, after many months' hurtling across the solar system, would be to arrive in the vicinity of their target planets, then skim past for just a few precious hours while snapping a couple of dozen fragile images, before once again hurtling off into the wastes of space. In the early years of rocket exploration, it was not always possible to slow down a craft and put it into orbit around such distant worlds as Mars or Venus. It was enough of a challenge simply to ensure that planetary probes could escape from the earth's gravity.

On July 15, 1965, Mariner IV flew past Mars after a trouble-free journey of 230 days, coming within about 6,000 miles of the planet at its closest approach. In these early days of remote-imaging technology, the transmission of pictures from deep space was no simple matter. Mariner's camera recorded 22 images onto a strip of photographic negative film, which was then processed internally on a convoluted series of rollers. A scanning device, similar in principle to an old-fashioned fax machine, then translated the results into radio signals for transmission to the waiting earth, with pulses corresponding to light or dark areas on the negatives.

Each picture took eight hours to transmit. In modern computer terms, this would represent a snail-like rate of eight bits per second. In order to send its entire batch of twenty-two images, Mariner's radio gear beamed away solidly for ten days, even as the spacecraft itself left Mars far behind and hurtled into deepest space. The original film's negative black-and-white values were then electronically reversed by the earth-based receiving equipment to produce positive images: the first close-up look at Mars in fuzzy frames of 200 scan lines, each consisting of a string of 200 tiny dots.

Today those images seem crude, but at the same time deeply touching, for so much effort was invested by so many people to achieve what we would now regard as so little reward. The merest fuzzy rumor of an image from deep space was once considered a technological miracle. Today we routinely expect high-resolution data from the farthest reaches of the solar system, and even from distant galaxies, via our space telescopes. Although we benefit from this perfection, some of the ghostly mystery of space has perhaps been stripped away by the searing clarity of our vision.

Opposite, ESA's proposal for a new and super-smart generation of Mars rovers has been approved for funding and development. ExoMars will be equipped with a biology lab to search for traces of life, past or present.

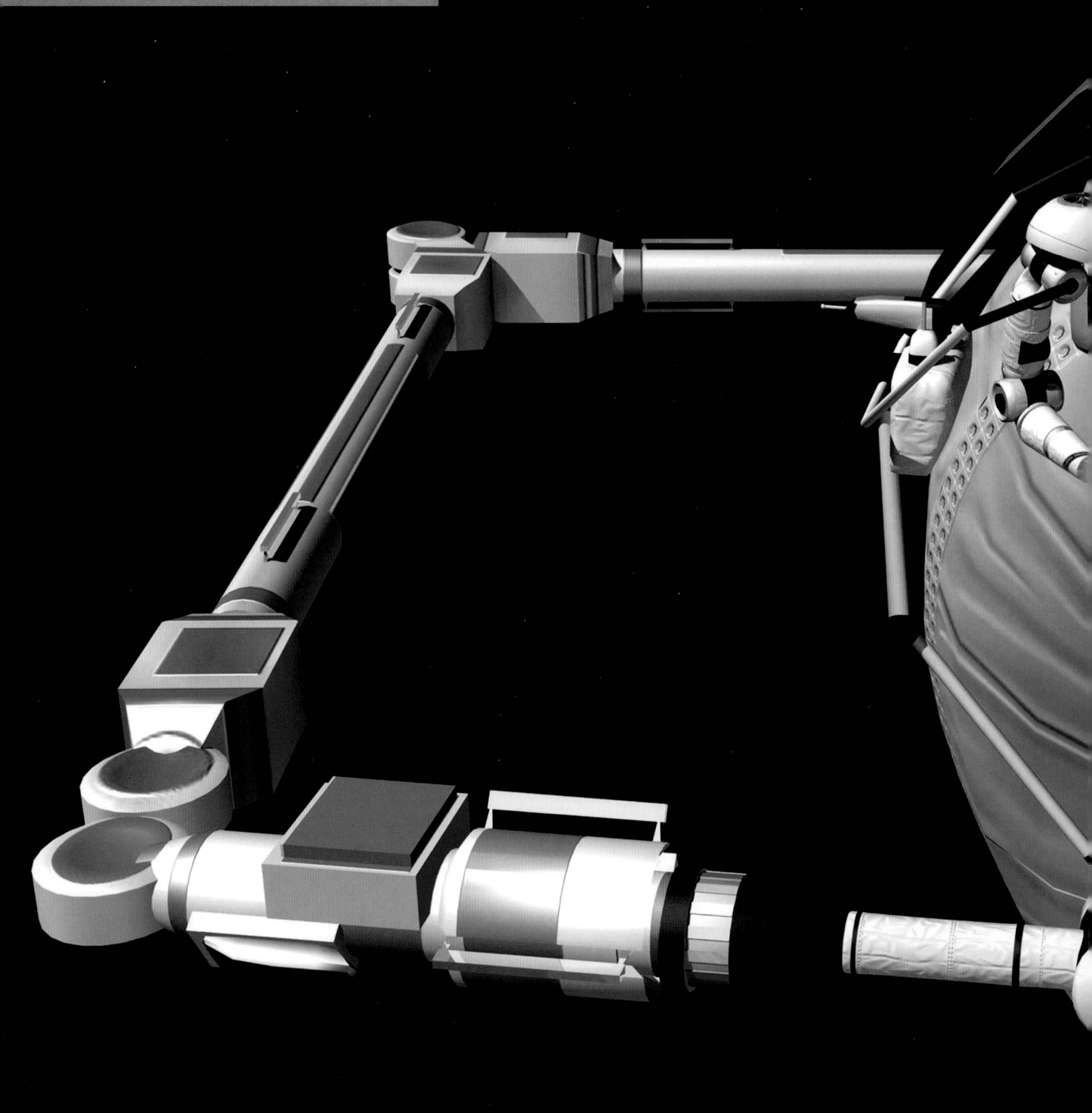

NASA is developing "Robonaut", a humanoid robotic tool to assist astronauts during space walks.

ROBONAUT
NASA

ROBONAUT
NASA

"One day we will not travel in spaceships. We will *be* spaceships."
Arthur C. Clarke, *Profiles of the Future*, 1973

Today's technology has made Mars seem almost as close and accessible as our own backyard. High-definition images of its surface can be downloaded on home computers within a few days or even hours of transmission. Ground controllers need time to process the images, but essentially they are made available to the public as soon as they are ready. We take for granted the god's eye views we now have of Mars, Jupiter, Saturn, dozens of moons and even several major asteroids.

The 1997 Mars Pathfinder mission opened a new era of internet space exploration. As the little wheeled rover Sojourner beamed back its startlingly clear images of the planet's surface, the relevant NASA websites received a world-record 33 million "hits" on July 4, touchdown day. Four days later that figure had risen to a total of 47 million. A small internet team (led by Kirk Goodall of the Jet Propulsion Laboratory in California) had revolutionized public awareness of planetary exploration on a budget so small it made almost no impact on Pathfinder's overall costs. We now expect to be able to log on and retrieve data from planetary rovers or deep space probes without a second thought. The visionaries of half a century ago imagined that we humans would visit the farthest reaches of the solar system. What they did not expect was that the solar system would be brought to us, without our so much as having to leave the house.

As Arthur C. Clarke has written: "One day we will not travel in spaceships. We will *be* spaceships." In a sense we already are. Robot probes are extensions of our minds and our physical reach, allowing us to scrabble in the soil of far-distant worlds with remotely operated claws. Does that mean, then, that we no longer have to go to the trouble of turning up in person to clutch handfuls of alien dust in our own hands? This has been the great debate ever since the Space Age began. Logic suggests that machine probes are the safest and most cost-efficient tools for space exploration. Instinct and emotion cause many of us to think differently.

Opposite, robots will extend our interactive exploration of space, but perhaps there is no true substitute for an actual human presence?

The greatest exploration

Above right, the Hubble Space Telescope has been described as NASA's most successful mission. It has captivated the public imagination with images such as this view of the Orion Nebula. Below right, a Spitzer Space Telescope image of Cassiopeia, the remnant of a supernova explosion.

> In 1960, just three years into the age of spaceflight, an experimental weather satellite called TIROS beamed back crude black-and-white television pictures of cloud tops and storms. Details were hard to distinguish, although the outlines of continents could be discerned, with practice, through unclouded windows in the atmosphere. One day an image analyst reported that he could just about make out faint patterns in snow on a picture of a region of northern Canada. At first his report was discounted, but it turned out that lumberjacks had indeed left large tracks in the snow at that very spot, dragging away logs from a forest clearing.

By 1961 the first astronauts and cosmonauts were peering at the earth through the tiny windows of their capsules. NASA astronaut Gordon Cooper, in his cramped Mercury ship, amazed mission controllers when he reported being able to see roads, buildings and even smoke from chimneys. By 1964 the clarity of the earth as seen from space could no longer be denied. Astronauts in two-man Gemini capsules corroborated each other's impressions and returned to earth with color photographs of drainage channels, wheat fields, and some interesting airfields and other man-made structures in eastern Europe and the Soviet Union.

These, were the kinds of exciting discoveries made by NASA and the civilian space effort in the first decade of the Space Age that were made public. Behind the scenes, however, a vast industry funded just as lavishly as the human space program was at work. It was operating in conditions of such secrecy that even the majority of politicians in Washington were not permitted to know about it. A series of "Discoverer" satellites was described in the press as being loaded with biological specimens and science experiments. In fact they contained Central Intelligence Agency (CIA) cameras.

Only in recent years has the extent of 1960s spy satellite technology been made clear. Three decades later and much of that technology has filtered into the civilian sector. We now study the earth from space more precisely than we do on the ground, with infrared, ultraviolet and radar wavelengths as well as in optical light. We take this multifaceted celestial view of ourselves and our home world more or less for granted today, but the surprise of those early TIROS technicians at the possibilities they had opened up reminds us that such a high vantage point has been available only in recent history.

And our lofty robot surveyors are showing us disturbing sights. Man's impact on the surface of the planet is impressive, yes, but also terrifying, for there is scarcely a scrap of land anywhere that does not carry his mark. Even the wildest mountain ranges and the murkiest depths of the sea betray chemical traces of human activity, carried on the winds or deposited by ocean currents. Space technology reveals our fragile condition in unprecedented detail, yet it may also help us learn how to protect the environment for future generations. This may yet prove to be the greatest prize that the space rocket has ever delivered.

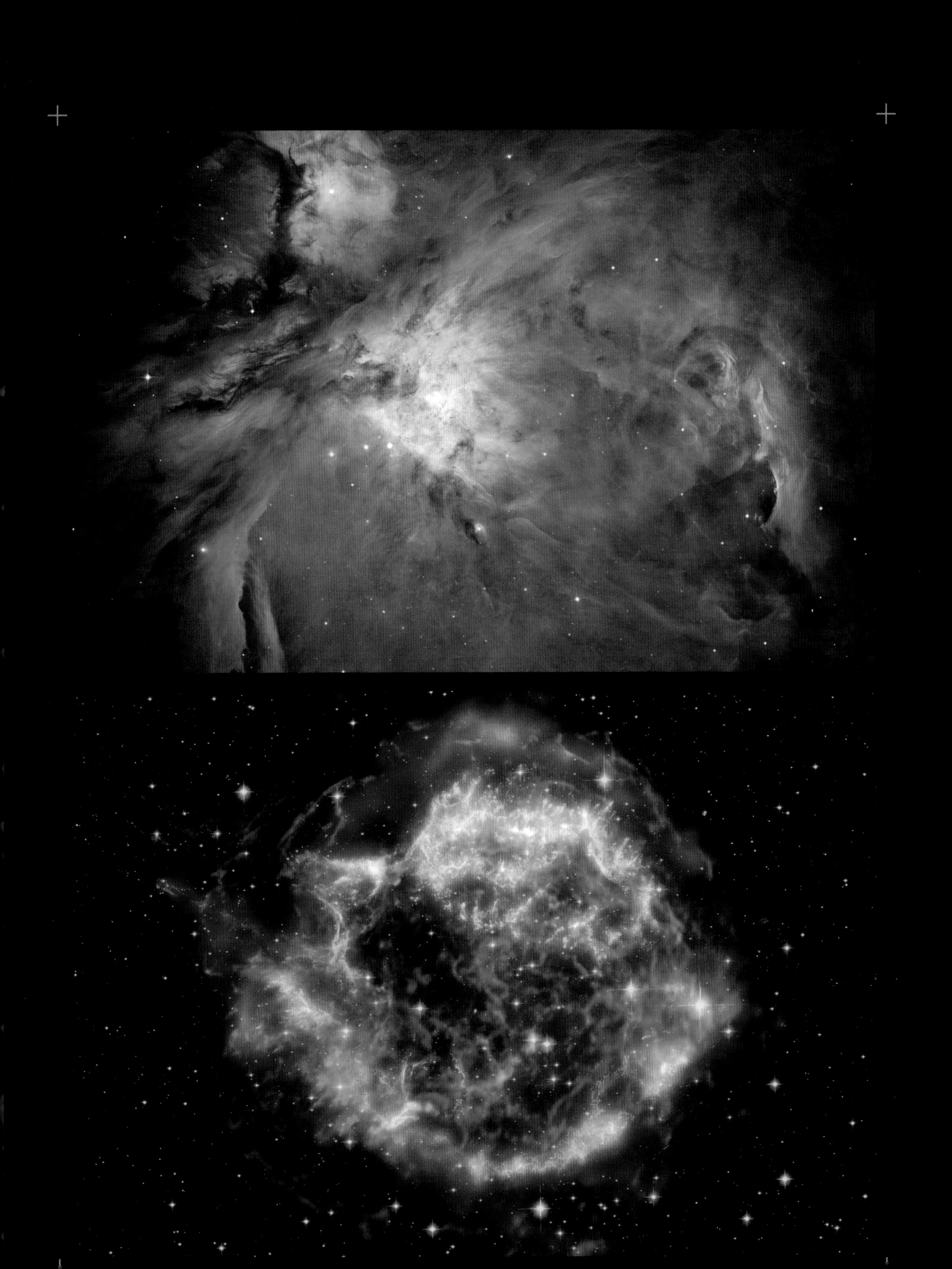

We shall not cease from exploration,
And the end of all our exploring,
Will be to arrive where we started,
And know the place for the first time.

T.S. Eliot, *Four Quartets*

Acknowledgments

At the Smithsonian National Air and Space Museum (NASM), Washington D.C., I would like to thank Dr Ted Maxwell for his support, along with the NASM's chief space historian, Roger Launius. Thanks also to my good friend the renowned space historian and archivist Frederick I. Ordway III, for so much help over the years. Randy Liebermann, an astute independent historian based near Washington D.C., also gave valuable advice and provided images, for which, many thanks. My appreciation also to Andy Aldrin, a wise commentator on Soviet space history.

On the digital front, Bob Sauls of John Frassanito & Associates was most welcoming to all my inquiries. He and his colleagues are among the finest computer graphics visualizers of NASA's future plans, and in particular, the next-generation lunar and Mars missions. Pat Rawlings, a very well-known space artist, was complimentary about this book project, and allowed us to use several of his artworks. Terry Sunday, an independent artist, made available his evocative modern-day computer graphics depiction of Wernher von Braun's early moonship proposal, while Dave Robinson gave us a wonderful image of a giant spacecraft heading towards a rendezvous with Mars.

Davide De Martin made available a number of fine images. I am also grateful to Kipp Teague and Eric Jones, whose archival work on the Apollo missions is widely admired by aerospace historians. Many other archivists and picture library personnel lent their expertise to this book. In particular, Margaret Persinger at the Kennedy Space Center, Mike Wright and Fred Deaton at the Marshall Space Flight Center, and Bill Ingalls at the Johnson Space Center, who preserves the work of an earlier generation of NASA photographers (such as the renowned Apollo-era photographer Bill Taub) as carefully as his own fine shots, in his capacity as senior NASA photographer. NASM photographers Dane Penland and Eric Long contributed samples of their very beautiful work, as did NASA chief photographer Bill Ingalls, and aerospace freelancer Mike Massee. In Europe, ESA's communications director Wendy Slater and her colleagues answered all my calls for data and images with extraordinary promptness.

Peter Tallack of the Conville & Walsh literary agency introduced me to Caz Hildebrand and her colleagues at the Here+There group. Many thanks to Caz, and to Mark Paton, for their hard and creative work designing and administering this project through to completion. Lily Richards, our principal picture researcher, worked many miracles, especially in connection with old Soviet materials. Thanks also to Martin Gledhill, and to the Hasselblad camera company (UK) for valuable support.

A number of interviewees contributed to the narratives in this book. Veteran cosmonaut Alexei Leonov, and the late Gherman Titov, told remarkable stories in 1997, while my friend and colleague Jamie Doran was preparing a groundbreaking film about Yuri Gagarin's life for BBC television; a film that subsequently led to a book written by Jamie and myself. It was a privilege to collect such vivid testimony from these authentic heroes of the early space age. Likewise, I have been fortunate enough over recent years to meet Buzz Aldrin, John Young, Dave Scott, Bill Pogue and many other veterans of NASA project Apollo. To all of them, my sincerest thanks. It has been a special privilege to talk with Dr Robert C. Seamans Jr., Associate Administrator of NASA from 1960 to 1965, and Deputy Administrator from 1965 to 1968.

Picture Credits

AKG-images: 22, 32, 39, 153
AP/EMPICS: 152, 179, 195
Adam Bartos: 114-115
Piers Bizony: 68, 69 Piers Bizony/Bonestell Art: 32, 157
ESA: 161, 169, 174, 175, 214, 215, 221, 224, 229, 231, 232, 233, 236, 237, 291(bottom), 301 (top and bottom), 304, 305, 306
ESA/Olivier de Goursac: 292-293 (top)
John Frassanito & Associates/Bob Sauls: 207, 256-257, 258-259, 260-261, 262-263, 264, 265, 266-267, 268, 279. 280, 282-283, 308-309, 310
Dennis Gilliam/MGM 158, 163, 273
Getty Images: 16, 52 right, 53
Bill Ingalls: 20-21, 24-25, 178, 182-183, 234-235
JAXA: 226-227
Magnum Photos/Jonas Bendiksen: 209
Mary Evans Picture Library: 18, 19 top and bottom, 288-289
Randy and Yulia Liebermann Lunar and Planetary Exploration Collection: 30 Frederic W. Freeman Estate: 31, 159
Mike Massee/Scaled Composites: 244, 246-247
NASA/DFRC: 84, 85, 86, 87 top & 87 bottom, 88, 89, 90, 91, 191
NASA/HSTI: 200, 203, 313 (top)
NASA/JPL: 36, 37, 287, 291 (top left & top right), 293 (bottom), 295, 296-297, 298-299, 302, 313 (bottom)
NASA/JSC: 2-3, 6, 45, 47, 55, 66 left & right, 67, 78, 79, 80-81, 82, 83, 96, 109, 125, 136, 137, 138 & 149 (all frames), 143, 144, 145-146, 147-148, 149, 150, 168 (all four frames), 171, 172-173, 180-181, 196-197, 198, 210-211 (all four frames), 212-213, 217, 218, 219, 222-223, 314-315
NASA/Kipp Teague: 108, 119, 121, 133, 134, 135, 255
NASA/KSC: 42, 43, 56, 58, 59, 74, 76, 99, 103, 104, 105, 107, 128-129, 130-131, 140, 141, 142, 184-185 (all four frames), 186, 187, 188-189, 192, 193, 194
NASA/MSFC: 13, 27, 28, 34, 35, 101, 165, 166
NASA/Bill Taub: 77, 95
NASA/Arden Wilfong: 65
Smithsonian Institution/National Air and Space Museum/Eric Long: 127 Smithsonian Institution/National Air and Space Museum/Dane Penland: 241
Pat Rawlings: 249, 252, 253
RIA Novosti: 40, 41, 51, 52 left, 61, 70-71, 72, 73
Dave Robinson: 276-277
Science and Society Picture Library: 62
Mark Shuttleworth Foundation: 243
Space Facts/Joachim Becker: 111, 122, 160 (all three frames)
Space Frontiers: 242, 250-251
Terry Sunday: 15
Time/Getty: 176, 201
XCORP/Mike Massee: 245

Index

Page numbers in *italic* refer to illustrations